Power Skills in Mathematics—I

Power Skills in Mathematics

Martine Rongier Wiznitzer
Carol Simeon Scheer

Project Editors

Marie Talbot
Instructor
Career Preparation Center
San Mateo County Office of Education
Daly City, California

Theodore Silveira
Department of English
San Francisco State University
San Francisco, California

A Trafalgar House Book————

McGraw-Hill Book Company

New York, St. Louis, San Francisco, Auckland, Bogotá, Düsseldorf, Johannesburg, London, Madrid, Mexico, Montreal, New Delhi, Panama, Paris, São Paulo, Singapore, Sydney, Tokyo, and Toronto

Portions of the material in this book were originally published under the title *The Pre-G.E.D. Basic Skills Series: Basic Mathematics Skills*

Trafalgar House Publishing, Inc.
145 East 52nd Street, New York, New York 10022

Printing: 2 3 4 5 6 7 8 9   Year: 0 1 2 3 4 5 6 7 8

Contents

Introduction

HOW TO USE THIS BOOK

This book will help you understand arithmetic. You can work in the book without any help from anybody. Just work at your own speed and follow all the instructions that you are given.

    1. Each chapter is separated into sections. Each section starts with a number. Sometimes the number is large and bold; sometimes it is a smaller number; and sometimes it is a number with a letter.

Read all the sections that start with a large bold number. In many sections, there are film strips. The film strips give you important information. Read them carefully. Here is an example of a film strip that shows you how to read a book.

Open the book
Read the first page
When you're finished, turn the page
Read the second page
Read the third page
Turn the page
Continue until the book is finished
Close the book

1.1. Test Yourself

When you see this title, you will have to answer a question or work out a problem to see if you understand the explanation that is given for it. If your answer is right, you should go directly to the next section that has a large bold number, skipping the section with the small number and a letter.

    If your answer is wrong or you don't know what to do, you should go to the section that has a small number, or a number with a letter.

1.1A. This section gives you more of an explanation of how to solve the problem. When you understand the explanation of the right answer, you should go to the next section that has a large bold number.

1.2. Practice

This section gives you more problems to work out. Check your answers at the end of the chapter before you go on to the next section. This will help you to find the exact spot in which you might be having trouble.

1.3. More Practice

At the end of some chapters, you will have to work out more problems like those you find as you read each chapter. The additional practice will help you make sure that you remember everything that you have learned in the chapter.

1.4. Word Problems

Word problems are part of everyday life, and they are a large part of any arithmetic test. This book will teach you how to solve word problems. There are word problems for you to solve at the end of each chapter. Answer grids are provided so you can mark the correct answer from the five choices you are given for each word problem.

2. EVALUATIONS

Each group of chapters is separated by an evaluation. These evaluations are designed to see how you are doing at each stage. They are really just extra practice sections. If you have done the practice sections in each chapter, the evaluations should be easy. Don't worry if you do not do well on the pretest. If you answered all the questions right, you would not need this book!

3. DO IT YOUR WAY

If you have already learned how to add, subtract, multiply, and divide a way that is different from the way we show you, stick with your own way. As long as you get the same answers, it doesn't matter if you use a different method. Arithmetic is done different ways all over the world. It is enough to know one way to do it.

4. WHEN YOU FINISH THE BOOK

After you have gone through this book, you will have a very good understanding of arithmetic and you should feel comfortable with word problems. Take it easy. Take it step by step. Enjoy the book!

Pretest

This pretest is to find out what you already know. Do the problems as well as you can, then check your answers.

1. 7 + 6 =

2.
$$\begin{array}{r} 35 \\ + 42 \\ \hline \end{array}$$

3.
$$\begin{array}{r} 394 \\ + 257 \\ \hline \end{array}$$

4.
$$\begin{array}{r} 8,952 \\ + 1,079 \\ \hline \end{array}$$

5. 8 − 2 =

6.
$$\begin{array}{r} 87 \\ - 26 \\ \hline \end{array}$$

7.
$$\begin{array}{r} 635 \\ - 489 \\ \hline \end{array}$$

8.
$$\begin{array}{r} 1,093 \\ - 695 \\ \hline \end{array}$$

9. 7 × 8 =

10.
$$\begin{array}{r} 253 \\ \times \ \ 6 \\ \hline \end{array}$$

11.
$$\begin{array}{r} 97 \\ \times 35 \\ \hline \end{array}$$

12.
$$\begin{array}{r} 293 \\ \times 159 \\ \hline \end{array}$$

13. 25 ÷ 5 =

14. 6) 845

15. 32) 536

16. 19) 437

17. 106)21,518

18. 438)1,389

19. ¾ × ⁸⁄₉

20. 2 2/5 × 3 1/3

21. 2/5 ÷ 6/10

22. 4 4/5 ÷ 1 2/10

23. 6/7 + 6/7

24. 5 4/6 + 2 5/6

25. 9/16 − 5/16

26. 4 1/8 − 1 3/8

27. 3 2/5 + 4 2/3

28. 7 1/4 − 2 2/6

29. .025 + .17

30.
$$\begin{array}{r} 4.62 \\ + 9.49 \\ \hline \end{array}$$

31. .38 − .19

1

32. 4.325
   − 1.769

33. 5.3
   × 2.6

34. .204
   × .34

35. 8) $\overline{26.16}$

36. .34) $\overline{7.82}$

37. 8% of 50 =

38. 10% of 300 =

39. 2 ft 10 in
   + 2 ft 5 in

40. 3 yd 1 ft 3 in
   −   1 ft 8 in

41. 4 lb 5 oz
   ×   7

42. 3) $\overline{3 \text{ lb } 9 \text{ oz}}$

43. 3 qt 1 pt
   +   1 pt

44. 1 gal 1 qt
   −   3 qt

45. 5 hr 23 min
   ×   5

46. 4) $\overline{8 \text{ hr } 32 \text{ min}}$

47. +5 − 8 =

48. −7 + 9 =

49. □ + 8 = 10

50. 26 − □ = 22

WORD PROBLEMS

Try these word problems. In each one, first decide if you must add, subtract, multiply, or divide.

51. In 1973, 13,841 ships went through the Panama Canal. In 1974, 14,033 ships went through the canal. How many ships passed through the canal in the two years?
(1) 27,874   **(2)** 27,984   **(3)** 27,994   **(4)** 27,864
(5) none of these

52. The batteries in a flashlight were guaranteed for 30 days. The batteries still worked after 65 days. How many days over the guarantee did the batteries still work?
(1) 95   **(2)** 35   **(3)** 30   **(4)** 24
(5) none of these

53. A stadium has 30,750 seats. If 27,345 people come to a game, how many seats remain empty?
(1) 3,405   **(2)** 2,485   **(3)** 3,515   **(4)** 3,400
(5) none of these

1.   1   2   3   4   5

2.   1   2   3   4   5

3.   1   2   3   4   5

54. Razor blades come in packages that contain 5 blades. How many razor blades are there in 17 packages?
    (1) 85   (2) 105   (3) 82   (4) 80
    (5) none of these

55. A sanitation truck holds 3,850 pounds of garbage. How many pounds of garbage do 23 trucks hold?
    (1) 98,500   (2) 88,000   (3) 91,000   (4) 80,550
    (5) none of these

56. Five families share equally a moving van that has 32,565 feet of space. How much space does each family have?
    (1) 6,513   (2) 7,613   (3) 6,520   (4) 6,512
    (5) none of these

57. There were 720 passengers traveling on a 15-car train. Each car had the same number of passengers. How many passengers were in each car?
    (1) 58   (2) 48   (3) 49   (4) 45
    (5) none of these

58. There are 48 boxes of cereal in a carton. If there are 1,248 boxes of cereal, how many cartons are needed to pack all the boxes of cereal?
    (1) 25   (2) 36   (3) 26   (4) 28
    (5) none of these

59. Boxer Rocky Tough won 3/4 of a million dollars from a fight. He paid 1/3 of that amount in taxes. What fraction of a million did he pay in taxes?
    (1) 1/4   (2) 1/2   (3) 1/3   (4) 3/4
    (5) none of these

60. It took 5 1/3 cups of flour to make 12 mini-loaves of French bread. How many cups would it take to make 1 loaf?
    (1) 4/9   (2) 7/8   (3) 1 1/5   (4) 1 5/8
    (5) none of these

61. Olga won 12/15 of her free-style dance contests. What fraction of her contests did she lose?
    (1) 12/15   (2) 3/5   (3) 1/5   (4) 4/5
    (5) none of these

62. A carpenter cut a 3 1/2 foot piece of wood from a piece of lumber that was 9 1/3 feet long. How many feet long was the piece that remained?
    (1) 6 5/6   (2) 5 5/6   (3) 6 1/3   (4) 6 1/2
    (5) none of these

4. 1 2 3 4 5

5. 1 2 3 4 5

6. 1 2 3 4 5

7. 1 2 3 4 5

8. 1 2 3 4 5

9. 1 2 3 4 5

10. 1 2 3 4 5

11. 1 2 3 4 5

12. 1 2 3 4 5

63. A truck brought a 3.5-ton load of sand to a building site. There were already 12.58 tons of sand there. How many tons of sand were there in all?
(1) 15.08   **(2)** 12.93   **(3)** 15.63   **(4)** 16.08
(5) none of these

    13.   1   2   3   4   5

64. For 1,500 kilowatthours of electricity per month a customer is billed $60.98 during the winter and $80.48 during the summer. How much higher is the electricity bill in the summer?
(1) $20.40   **(2)** $20.58   **(3)** $19.00   **(4)** $19.50
(5) none of these

    14.   1   2   3   4   5

65. Carla bought five towels for $1.98 each. What was the total price of the towels?
(1) $10.00   **(2)** $9.98   **(3)** $9.50   **(4)** $9.90
(5) none of these

    15.   1   2   3   4   5

66. A road 15.25 miles long has a marker every .05 mile. How many signs are there?
(1) 30.5   **(2)** 305   **(3)** 350   **(4)** 3.05
(5) none of these

    16.   1   2   3   4   5

67. Your taxable income is $8,750. Your personal exemption credit is 2 percent of your taxable income. How much is your exemption credit?
(1) $8,752   **(2)** $1,758   **(3)** $750   **(4)** $175
(5) none of these

    17.   1   2   3   4   5

68. You put a 1 ft 8 in lamp on top of a 2 ft 6 in high table. How far is it from the floor to the top of the lamp?
(1) 2 yd 16 in   **(2)** 3 ft 18 in   **(3)** 1 yd 1 ft 2 in
(4) 3 ft 1 in   **(5)** none of these

    18.   1   2   3   4   5

69. You carry home a 3 pound 12 ounce box of detergent, 1 pound of hot dogs, and a 1 pound 2 ounce cake mix. How many pounds of groceries do you carry in all?
(1) 6 lb   **(2)** 5 lb 10 oz   **(3)** 5 lb 15 oz
(4) 5 lb 14 oz   **(5)** none of these

    19.   1   2   3   4   5

70. A roofer worked 1 week and 3 days on the Lee's house, 6 days on the McBride's home, and 1 week and 5 days on the DiPalma's house. How long did it take the roofer to finish all three jobs?
(1) 4 weeks   **(2)** 3 weeks and 6 days
(3) 3 weeks and 10 days   **(4)** 2 weeks and 11 days
(5) none of these

    20.   1   2   3   4   5

71. It was 6°F at 8 P.M. During the night the temperature drops 12 degrees. What is the morning temperature?
  (1) +18°F   **(2)** –14°F   **(3)** –6°F   **(4)** –18°F
  (5) none of these

21. 1 2 3 4 5

72. What number added to 13 equals 18?
  (1) 6   **(2)** 5   **(3)** 31   **(4)** 15
  (5) none of these

22. 1 2 3 4 5

73. According to the circle graph, what kind of vehicles are seen the MOST on I-88?
  (1) jeeps   **(2)** cars   **(3)** trucks   **(4)** campers
  (5) none of these

23. 1 2 3 4 5

Vehicles on I-88

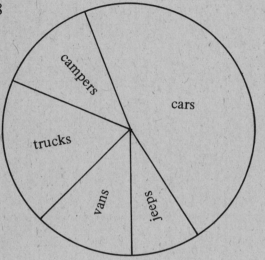

PRETEST ANSWERS

(1) 13	**(2)** 77	**(3)** 651	**(4)** 10,031	**(5)** 6
(6) 61	**(7)** 146	**(8)** 398	**(9)** 56	**(10)** 1,518
(11) 3,395	**(12)** 46,587	**(13)** 5	**(14)** 140r5	**(15)** 16r24
(16) 23	**(17)** 203	**(18)** 3r75	**(19)** 2/3	**(20)** 8

(21) 2/3   **(22)** 4   **(23)** $1\frac{7}{7} = 1\frac{5}{7}$ **(24)** 7 9/6 = 8 1/2   **(25)** 4/16 = 1/4

(26) 2 6/8 = 2 3/4   **(27)** 8 1/15   **(28)** 4 11/12   **(29)** .195   **(30)** 14.11

(31) .19   **(32)** 2.556   **(33)** 13.78   **(34)** .06936   **(35)** 3.27

(36) 23   **(37)** 4   **(38)** 30   **(39)** 4 ft 15 in = 1 yd 2 ft 3 in

(40) 2 yd 2 ft 7 in   **(41)** 28 lb 35 oz = 30 lb 3 oz   **(42)** 1 lb 3 oz

(43) 4 qt = 1 gal   **(44)** 2 qt   **(45)** 25 hr 115 min = 1 day 2 hr 55 min

(46) 2 hr 8 min   **(47)** –3   **(48)** +2   **(49)** 2   **(50)** 4

WORD PROBLEMS

51. (1)

$$\begin{array}{r} 13,841 \\ + \ 14,033 \\ \hline 27,874 \end{array}$$

52. (2)

$$\begin{array}{r} 65 \\ - \ 30 \\ \hline 35 \end{array}$$

53. (1)

$$\begin{array}{r} 30,750 \\ - \ 27,345 \\ \hline 3,405 \end{array}$$

54. (1)

$$\begin{array}{r} 17 \\ \times \ 5 \\ \hline 85 \end{array}$$

55. (5)

$$\begin{array}{r} 3,850 \\ \times \quad 23 \\ \hline 11\ 550 \\ 77\ 000 \\ \hline 88,550 \end{array}$$

56. (1)

$$5\ )\ \overline{32,565} \quad 6,513$$

57. (2)

$$\begin{array}{r} 48 \\ 15\ )\ \overline{720} \\ 60 \\ \hline 120 \\ 120 \end{array}$$

58. (3)

$$\begin{array}{r} 26 \\ 48\ )\ \overline{1,248} \\ 96 \\ \hline 288 \\ 288 \end{array}$$

59. (1) $\dfrac{1}{3} \times \dfrac{3}{4} = \dfrac{1}{\cancel{3}} \times \dfrac{\overset{1}{\cancel{3}}}{4} = \dfrac{1}{4}$

60. (1) $5\ 1/3 \div 12 = \dfrac{\overset{4}{\cancel{16}}}{3} \times \dfrac{1}{\underset{3}{\cancel{12}}} = \dfrac{4}{9}$

61. (3)

$$\begin{array}{r} 15/15 \\ - \ 12/15 \\ \hline 3/15 = 1/5 \end{array}$$

62. (2)

$$\begin{array}{r} 9\ 1/3 = 9\ 2/6 = 8\ 8/6 \\ - \ 3\ 1/2 = 3\ 3/6 = 3\ 3/6 \\ \hline 5\ 5/6 \end{array}$$

63. (4)

$$\begin{array}{r} 12.58 \\ + \ 3.5 \\ \hline 16.08 \end{array}$$

64. (4)

$$\begin{array}{r} \$80.48 \\ - \ 60.98 \\ \hline \$19.50 \end{array}$$

65. (4)

$$\begin{array}{r} \$1.98 \\ \times \quad 5 \\ \hline \$9.90 \end{array}$$

66. (2)

$$.05.\ )\ \overline{15.25.} \quad 305.$$

67. (4)

$$\begin{array}{r} \$8,750 \\ \times \quad .02 \\ \hline \$175.00 \end{array}$$

68. (3)

$$\begin{array}{r} 1\ \text{ft}\ \ 8\ \text{in} \\ + \ 2\ \text{ft}\ \ 6\ \text{in} \\ \hline 3\ \text{ft}\ 14\ \text{in} = 1\ \text{yd}\ 1\ \text{ft}\ 2\ \text{in} \end{array}$$

69. (4)

$$\begin{array}{r} 3\ \text{lb}\ 12\ \text{oz} \\ 1\ \text{lb} \\ + \ 1\ \text{lb}\ \ 2\ \text{oz} \\ \hline 5\ \text{lb}\ 14\ \text{oz} \end{array}$$

70. (1)

$$\begin{array}{l} 1\ \text{week}\ 3\ \text{days} \\ \qquad\qquad 6\ \text{days} \\ + \ 1\ \text{week}\ 5\ \text{days} \\ \hline 2\ \text{weeks}\ 14\ \text{days} = 4\ \text{weeks} \end{array}$$

1 week = 7 days

71. (3) +6 – 12 = –6

72. (2) 13 + □ = 18
13 + □ – 13 = 18 – 13
□ = 5
13 + ⑤ = 18 o.k.

73. (2) cars

ONE

Numbers

1. DIGITS AND NUMBERS

The digits 0, 1, 2, 3, 4, 5, 6, 7, 8, 9 are used to make numbers. They tell "how many." You can talk about very large amounts or very small amounts using only these digits.

1.1. With the digits 3 and 6 you can make the number 36. But if you put the digits the other way you get 63. 63 is not the same number as 36. The PLACE in which you put a digit is important.

1.2. PLACE VALUE means that putting a digit in one spot makes that digit worth something different from what it would be worth if it were in another spot. Here are the places for the digits:

millions	hundred thousands	ten thousands	thousands	hundreds	tens	units
————	————	————	————	————	————	————

The number 36 has 6 units. The number 63 has 3 units.

1.3. Learn the name of each place so that you will be able to tell how many units, tens, hundreds, thousands and so on are in any number.

1.4. Test Yourself

(a) How many tens in 87?

(b) How many thousands in 5,659?

Check your answers. If they are right, go to section 1.5. If they are wrong or if you don't know what to do, go to section 1.4A.

1.4A. Using the place value chart in section 1.2, put the digits in the spaces starting from the right side, the units place.

hundreds	tens	units	
————	8	7	87 has 8 tens

Do the same for the next problem:

hundred thousands	ten thousands	thousands	hundreds	tens	units	
————	————	5	6	5	9	5,659 has 5 thousands.

Go to section 1.5.

1.5 Practice

Find how many units there are in the following:

    (a) 43   **(b)** 68   **(c)** 104   **(d)** 9,853

Find how many tens there are in the following:

    (e) 67   **(f)** 143   **(g)** 76   **(h)** 34

Say how many units, tens, hundreds, and thousands there are in each number:

    (i) 43   **(j)** 3,658   **(k)** 863   **(l)** 259   **(m)** 6,952

Check your answers.

2. WRITING NUMBERS

Writing numbers is easy because each place has a name and one digit goes in each place. Here are some examples:

	TTh	Th	H	T	U
sixty-five				6	5
four hundred thirty-nine			4	3	9
eight thousand, three hundred twenty-one		8,	3	2	1
thirty-two thousand, seven hundred ninety-four	3	2,	7	9	4

2.1. Test Yourself

Write with digits: six thousand, five hundred forty-three. Check your answer. If it is right go to section 2.2. If it is wrong or if you don't know what to do, go to section 2.1A.
2.1A. Look back at the place value chart in section 1.2. Put 6 in the thousands place. Fill in each place with a digit.

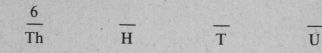

The digit 5 goes in the hundreds place. You can finish this and go to section 2.2.

2.2. Practice

Write with digits:
(a) fifty-eight
(b) nine hundred fifty-three
(c) four thousand, seven hundred twenty-one

(d) thirteen thousand, six hundred seventy-nine

(e) two hundred thirty-four thousand, five hundred eighty-seven

Check your answers.

3. RESTING TO READ BETTER

When you write a number that has 4 digits or more, always put a comma three digits from the right:

26,758   twenty-six thousand, seven hundred fifty-eight

2,643   two thousand, six hundred forty-three

Commas make it easier to read large numbers.

Commas separate the thousands from the hundreds:

ten thousands   thousands,   hundreds   tens   units

3.1. Practice

Put the commas in the right places and read these large numbers. It helps to read out loud.

(a) The Brooklyn Battery Tunnel, 9117 feet long, is the longest underwater tunnel in the U.S.

(b) Detroit's Tiger Stadium seats 54226 people.

(c) Three years ago there were 39064 cocker spaniel dogs registered with the American Kennel Club.

(d) 9564 athletes took part in the Montreal Olympics.

(e) The Amazon River is 3915 miles long.

Check your answers.

4. ZERO IS A DIGIT

Look at these prices:

$ 8,004   eight thousand, four dollars

$ 804   eight hundred four dollars

$ 84   eighty-four dollars

Which is the cheapest? Which is the most expensive?

When it comes to money, zeros make a great difference! In arithmetic, learning to place zeros correctly is very important.

Here are some more examples:

7,032   seven thousand, thirty-two

70,320   seventy thousand, three hundred twenty

4.1. Test Yourself

(a) Read this number: 13,093

(b) Write this number with digits: seven thousand thirty-nine

Check your answers. If they are right, go to section 4.2. If they are wrong or if you don't know what to do, go to section 4.1A.

4.1A. If you have problems reading the number, make columns: the units column on the right, then the tens column, and so on:

ten thousands	thousands	hundreds	tens	units

Put your digits in the columns, starting from the right:

ten thousands	thousands	hundreds	tens	units
1	3,	0	9	3

Say what you see in your columns, stopping at the comma and not saying the zero:

    thirteen thousand ninety-three

To write the number:

    seven thousand thirty-nine

with digits, do the same thing. Make your columns:

thousands	hundreds	tens	units

Put the digits in the columns:

thousands	hundreds	tens	units
7		3	9

The hundreds column is empty: you need a zero to hold the place:

thousands	hundreds	tens	units
7	0	3	9

Put a comma three digits from the right. Your answer is 7,039. If you got 739 instead, you probably forgot to put in the zero to hold the hundreds place.

4.2. Practice

Read these numbers:

    (a) 6,039   (b) 74,600   (c) 32,908   (d) 10,030

Write with digits:

    (e) six thousand, thirty-two

    (f) two hundred five

    (g) twenty-two thousand, three hundred thirty

When you write a check, the amount must be written out.

Here is a check:

```
                                                              953
     MICHELE JOAN COLE

                          June 2  19 79        51-7013
                                               2111
PAY TO THE
ORDER OF    Phyliss Smith                   $364 79/100

     Three hundred sixty four and 79/100 —— DOLLARS

          ■┐ NEW CITY BANK
          └┴ OF AMERICA
             NEW YORK, NEW YORK    Michele Joan Cole
FOR
 ⑆2111⑈7013⑆  505 00572 0⑈ 09
```

You need to know how to write the words for numbers if you write checks.

Write with words:
    (h) 546   **(i)** 72   **(j)** 3,589

5. NUMBER REMINDERS

Digits 0, 1, 2, 3, 4, 5, 6, 7, 8, 9 are used to make NUMBERS. The PLACE in which you put a digit is important. Here are some of the places you can put digits:

ten thousands	thousands	hundreds	tens	units

        4 2 9 has 9 UNITS

        4 9 2 has 9 TENS

        9 4 2 has 9 HUNDREDS

In 5,642 there are 5 thousands, 6 hundreds, 4 tens, and 2 units. To read this number say: five thousand, six hundred forty-two.

Starting at the units place and moving to the left, put a comma after 3 digits. This makes it easier to read larger numbers. 469,253

705 is different from 7,005. In arithmetic, learning to place zeros correctly is very important.

You need to know how to write the words for numbers if you write checks. 289 written out is two hundred eighty-nine.

6. MORE PRACTICE

6.1. Say how many units there are in each number:
(a) 704   (b) 63   (c) 630   (d) 98   (e) 56

6.2. Say how many hundreds there are in each number:
(a) 501   (b) 953   (c) 876   (d) 7,335   (e) 4,427

6.3. Read these numbers:
(a) 58   (b) 804   (c) 2,006   (d) 7,010

6.4. Write these numbers with digits (don't forget the commas):
(a) four thousand, six   (b) three hundred five   (c) eight thousand, thirty

6.5. Write with words:
(a) 89   (b) 703   (c) 4,019

Check your answers.

If they are all right, start Chapter 2. If one or two answers are wrong, make sure you understand the correct answers and go to Chapter 2. If more than two answers are wrong, read Chapter 1 again. Then go to Chapter 2.

7. ANSWERS

1.4. (a) 8 TENS (b) 5 THOUSANDS

1.5. (a) 3 units   (b) 8 units   (c) 4 units   (d) 3 units
(e) 6 tens   (f) 4 tens   (g) 7 tens   (h) 3 tens

	Th	H	T	U
(i)			4	3
(j)	3	6	5	8
(k)		8	6	3
(l)		2	5	9
(m)	6	9	5	2

2.1. 6,543

2.2. (a) 58   (b) 953   (c) 4,721   (d) 13,679   (e) 234,587

3.1. **(a)** 9,117   (nine thousand, one hundred seventeen)
    (b) 54,226   (fifty-four thousand, two hundred twenty-six)
    (c) 39,064   (thirty-nine thousand, sixty-four)
    (d) 9,564   (nine thousand, five hundred sixty-four)
    (e) 3,915   (three thousand, nine hundred fifteen)

4.1. **(a)** thirteen thousand, ninety-three   **(b)** 7,039

4.2. Say:
    (a) six thousand, thirty-nine   **(b)** seventy-four thousand, six hundred
    (c) thirty-two thousand, nine hundred eight   **(d)** ten thousand, thirty
    Write:
    (e) 6,032   **(f)** 205   **(g)** 22,330   **(h)** five hundred forty-six
    (i) seventy-two   **(j)** three thousand, five hundred eighty-nine

6.1. **(a)** 4   **(b)** 3   **(c)** 0   **(d)** 8   **(e)** 6

6.2. **(a)** 5   **(b)** 9   **(c)** 8   **(d)** 3   **(e)** 4

6.3. Say:
    (a) fifty-eight   **(b)** eight hundred four
    (c) two thousand, six   **(d)** seven thousand, ten

6.4. **(a)** 4,006   **(b)** 305   **(c)** 8,030

6.5. **(a)** eighty-nine   **(b)** seven hundred three
    (c) four thousand, nineteen

TWO

Addition: Part I

1. WHAT IS ADDITION?

Joining groups of things to find their total amount is called ADDITION. Adding numbers together answers the question: "How many *in all?*"
Here is an example:

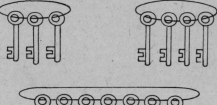

How many keys in all?

There are two ways of writing this addition.
You can write it either this way:

   3 + 4 = 7

or this way:

$$\begin{array}{r} 3 \\ + 4 \\ \hline 7 \end{array}$$

To read this addition you say: 3 plus 4 is 7.

1.1. Test Yourself

Here is another example of an addition:

How many hangers in all?
Check your answer. If it is right, go to section 1.2. If it is wrong, or if you don't know what to do, go to section 1.1A.

1.1A. If you put the hangers together, you will have:

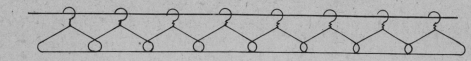

Go to section 1.2.

1.2. The two ways of writing this addition are:

2 + 6 = 8   or   2
                       + 6
                       ———
                         8

2. THE ADDITION CHART

If you learn the basic addition facts, you'll be able to add more quickly. The addition chart shows the basic facts all at once.

Here is how to use the addition chart: For example, let's add 2 + 5.

Find the number 2 in the side row

Go straight over the row and stop under the 5

Your answer is in the box where you stopped

Write it down:
2 + 5 = 7   or   2
                + 5
                ——
                 7

+	0	1	2	3	4	5	6	7	8	9
0	0	1	2	3	4	5	6	7	8	9
1	1	2	3	4	5	6	7	8	9	10
2	2	3	4	5	6	7	8	9	10	11
3	3	4	5	6	7	8	9	10	11	12
4	4	5	6	7	8	9	10	11	12	13
5	5	6	7	8	9	10	11	12	13	14
6	6	7	8	9	10	11	12	13	14	15
7	7	8	9	10	11	12	13	14	15	16
8	8	9	10	11	12	13	14	15	16	17
9	9	10	11	12	13	14	15	16	17	18

Notice that zero plus any number is always that number.

2.1. Test Yourself

Use the addition chart to find the answer for 6 + 3.
Check your answer.
If it is right, go to section 3.
If it is wrong, or if you don't know what to do, go to section 2.1A.
2.1A. To find this answer, follow the same directions as before. Find the number 6 in the side row. Go straight over the row and stop under the 3 (be careful to stay in the same row). Your answer is in the box where you stopped: 6 + 3 = 9   or

$$\begin{array}{r} 6 \\ + \ 3 \\ \hline 9 \end{array}$$

3. HELPFUL HINTS FOR ADDITION

The PLUS SIGN is used for addition. It tells you to ADD the numbers you see. It looks like this: +

When you add, there are two ways of writing your answer:

4 + 5 = 9   or

$$\begin{array}{r} 4 \\ + \ 5 \\ \hline 9 \end{array}$$

The answer in addition is called the SUM.
To check your addition, here is all you do:

$$\begin{array}{r} 4 \\ + \ 5 \\ \hline 9 \end{array}$$ Add down

$$\begin{array}{r} 4 \\ + \ 5 \\ \hline \end{array}$$ Then add up to check

Same answer?
You got it right.

3.1. Practice

Add these numbers. Use the addition chart only if you need to.

   (a) $\begin{array}{r} 4 \\ + \ 2 \\ \hline \end{array}$ **(b)** $\begin{array}{r} 3 \\ + \ 0 \\ \hline \end{array}$ **(c)** $\begin{array}{r} 7 \\ + \ 1 \\ \hline \end{array}$ **(d)** $\begin{array}{r} 5 \\ + \ 2 \\ \hline \end{array}$

   (e) 2 + 3 =   **(f)** 0 + 8 =

   (g) 6 + 5 =   **(h)** 3 + 9 =

Check your answers. If all your answers are right, go to section 4. If one or more are wrong, read sections 1 to 2 again before you go on.

4. ADDING TWO-DIGIT NUMBERS

A clerk has 13 green umbrellas and 15 striped umbrellas. How many umbrellas does the clerk have to sell in all?

To add a two-digit number to another two-digit number, use the column form. Put units under units, and tens under tens.

Add the units column:
$$\begin{array}{r} 3 \\ + 5 \\ \hline 8 \end{array}$$

Then add the tens column:
$$\begin{array}{r} 1 \\ + 1 \\ \hline 2 \end{array}$$

Your answer is 28.
The clerk has 28 umbrellas to sell.

tens column	units column
1	3
+ 1	5

tens column	units column
1	3
+ 1	5
	8

tens column	units column
1	3
+ 1	5
2	8

4.1. Sometimes you don't need to add in the tens column. Here is an example:
$$\begin{array}{r} 63 \\ + 5 \\ \hline \end{array}$$

Add the units column

In the tens column, there is nothing to add to the 6; just bring it down.

Your answer is 68.

tens column	units column
6	3
+	5
	8

tens column	units column
6	3
+	5
6	8

4.2 Test Yourself

Do these additions:
$$\begin{array}{r} 63 \\ + 24 \\ \hline \end{array} \qquad \begin{array}{r} 32 \\ + 5 \\ \hline \end{array}$$

Check your answers: If they are right, go directly to 4.3. If one is wrong or if you don't know what to do, go to 4.2A.

4.2A. Before you start the addition, be sure to separate the numbers into two columns: the units column and the tens column. Add the units column, then add the tens column.

```
tens  units              tens  units
  6     3                   3     2
+ 2     4                +        5
─────────               ─────────
  8     7                   3     7
```

4.3. Practice

(a) 78   (b) 45   (c) 62   (d) 15   (e) 54
   + 21   + 42   + 27   + 23   + 3
   ────   ────   ────   ────   ────

Check your answers. If all your answers are right, go to section 4.4. If one or more are wrong, read sections 3 and 4 again.

4.4 An addition is not always given directly in the column form.
For example: 27 + 32 =

Put the
addition
in column
form.

```
tens   units
  2      7
+ 3      2
──────────
```

Add units
column.
Add tens
column.

```
tens   units
  2      7
+ 3      2
──────────
  5      9
```

Your answer is 59.

4.5. Practice

Put these additions in column form and do them:
(a) 34 + 52 =   (b) 26 + 31 =   (c) 57 + 40 =

Check your answers. If all your answers are right, go to section 5. If one or more are wrong, read section 4 again.

5. ADDING THREE–DIGIT NUMBERS

A postal clerk sold 236 stamps in the morning and 142 stamps in the afternoon. How many stamps did the clerk sell in all?
The addition is: 236 + 142 =

To add three-digit numbers, use the column form again. Put units under units, tens under tens, and hundreds under hundreds.
Add the units column.
Then add the tens column.
Then add the hundreds column.
Your answer is 378.
The clerk sold 378 stamps in all.

hundreds	tens	units
2	3	6
+ 1	4	2
3	7	8

5.1. Test Yourself

Do this addition: 124 + 541 =

Check your answer. If it is right, go to section 5.2. If it is wrong, or if you don't know what to do, read section 5 again.

5.2. Practice

Do these additions. If they are not in column form, be sure to put them in column form first.

(a)   456   (b)   382   (c)   625   (d)   237   (e) 347 + 421 =
     + 123   + 207   + 222   + 42   (f) 192 + 705 =
                                                                                (g) 528 + 231 =
                                                                                (h) 263 + 25 =

Check your answers.

6. ADDING MORE THAN TWO NUMBERS

Try to add these numbers all at once:
        4
        2
      + 3
     ————

Can't do it? No. It's impossible to add more than two numbers at a time! Here is what you have to do:

Add the first two numbers.

Now add the sum to the last number.

Your answer is 9.

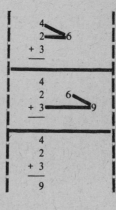

6.1. Test Yourself

Do this addition: 5
                             1
                        + 2
                        ———

Check your answer. If it is right, go to section 6.1A. If not, read section 6 again.

6.1A. You can add only two numbers at a time.

Add the 5 and the 1.   5
You get 6.   1——6
                    + 2
                    ———

Next add the 6 to the 2   5
                      1   6
                  + 2——8

Your answer is 8.
Go to section 6.2.

6.2. Practice

  (a)   3   **(b)**   2   **(c)**   6   **(d)**   5   **(e)**   4
           2   4   1   3   1
       + 4   + 2   + 2   + 0   + 2
       ———   ———   ———   ———   ———

Check your answers. If they are all correct, go to section 6.3. If not, read section 6 again.

6.3. Practice

  (a)   2   **(b)**   3   **(c)**   1   **(d)**   3   **(e)** 0 + 1 + 4 + 4 =
           1   3   2   2
           1   1   2   1
       + 4   + 2   + 2   + 3
       ———   ———   ———   ———

Check your answers.

7. ADDING SEVERAL TWO-DIGIT NUMBERS

Example: 23
                42
          + 14
          ———

Add the units column first, adding only two numbers at a time.

Then, add the tens column the same way.

Your answer is 79.

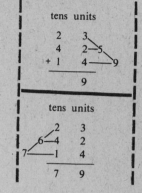

7.1. Practice

(a)	22	(b)	63	(c)	45	(d)	14	(e)	10
	35		11		21		30		24
	+ 12		+ 14		+ 20		12		32
							+ 23		10
									+ 13

Check your answers.

8. MORE PRACTICE

Do these additions. When an addition is not in the column form, put it in the column form first. When you need to add several numbers, always add two numbers at a time.

(a) 9 + 9 =   (b)   41   (c)   871   (d)   64   (e)   7
                  + 36   + 107   + 23   + 6

Check your answers.

9. WORD PROBLEMS—ADDITION

ADDING is joining groups of things to get their total amount. Addition problems are easier to solve if you look for addition word clues. Some addition word clues are:

(a) ALTOGETHER—The team scored 3 touchdowns in the first quarter, none in the second, 2 in the third quarter, and 3 in the last quarter. How many touchdowns were scored altogether?

(b) INCREASED—Coffee cost 3 dollars a pound last month. The price increased 1 dollar this month. What is the price of coffee this month?

(c) TOTAL—A shopper bought 12 cans of soup, 4 loaves of bread, and 2 kinds of meat. What was the total number of items purchased?

(d) IN ALL— Twelve boxes are in the closet. Thirty-two are on the shelf. How many boxes are there in all?

(e) GAINED—A dog weighed 43 pounds but it gained 12 extra pounds when it was fed too much. How much does the dog weigh now?

(f) MORE—Twenty-four people work in the records department at city hall. Thirteen more people are hired for the summer. How many people are now on the payroll?

(g) SUM—What is the sum of 42 + 13 + 24?

10. HOW TO SOLVE WORD PROBLEMS

Here is how to do the first word problem. We give you the explanations step by step. Do all your word problems the same way.

10.1. Read the Problem Carefully

Don't worry if there are words you don't understand. You can do this problem even if you don't know what a touchdown is.

> Problem: The team scored 3 touchdowns the first quarter, 0 the second, 2 the third, and 3 the last quarter. How many touchdowns were there altogether?

10.2. What Must You Find Out?

Look for words that ask you to find something.

> You must find out:
>
> How many touchdowns were there altogether?

10.3. Look for Word Clues

Word clues will tell you what you have to do. *Altogether* means that you must add.

> How many touchdowns were there ALTOGETHER?

10.4. Get Rid of Useless Information

Word problems may have a lot of information that you don't need. To make sure you understand the problem, say it in your own words.

> The problem in your own words:
>
> There are 3 touchdowns, then 0 touchdowns, then 2 touchdowns, then 3 touchdowns.
>
> How many are there altogether?

10.5. Write a Number Sentence

A number sentence uses numbers, not words. Once you understand the problem, write it with numbers only.

10.6. Solve the Problem

10.7. Check the Answer

Make sure that the answer you got is the right one.

10.8. Give Your Answer

Find the choice that matches yours. Fill in the column below.

> The number sentence for this problem:
>
> $3 + 0 + 2 + 3 =$
>
> ---
>
> $3 + 0 + 2 + 3 = 8$
>
> ---
>
> Do all touchdowns add up to 8? If yes, you got it right.

11. ADDITION WORD PROBLEMS

Follow the same steps to solve all these problems. At the end of each problem there are five choices for you to pick from. These are multiple choice answers. For best results, you should try not to look at the answers given until you have found the answer yourself. When you have your answer, look at the answers given and find the choice that matches yours. If none of the answers matches yours, your answer should be "none of these."

Fill in the column under the number that is your answer.

Example:

Your answer is 4. The choices are

    (1) 2   **(2)** 3   **(3)** 0   **(4)** 4

    (5) none of these

You must answer:

```
1   2   3   4   5
⋮   ⋮   ⋮   █   ⋮
```

Example:

Your answer is 2. The choices are

    (1) 5   **(2)** 3   **(3)** 1   **(4)** 7

    (5) none of these

You must answer:

```
1   2   3   4   5
⋮   ⋮   ⋮   ⋮   █
```

Now try these word problems:

1. A group of 327 people came to the bazaar in the morning. Another 562 people came in the afternoon. What was the total attendance at the bazaar?

    (1) 988   **(2)** 890   **(3)** 889   **(4)** 789
    (5) none of these

    1.   1   2   3   4   5

2. The fire department received 12 calls during the daytime and 4 calls at night. What was the total number of calls the fire department received in 24 hours?

    (1) 36   **(2)** 16   **(3)** 28   **(4)** 15
    (5) none of these

    2.   1   2   3   4   5

3. You have $45 in your savings account. You deposit $33. How much money do you have now?

    (1) $68   **(2)** $77   **(3)** $78   **(4)** $88
    (5) none of these

    3.   1   2   3   4   5

4. At softball practice, Lisa hit the ball 2 times her first time up, 4 times the second time, 1 time her third try, and 2 times her last time up. How many times in all did Lisa hit the ball during softball practice?

    (1) 7   **(2)** 5   **(3)** 8   **(4)** 9
    (5) none of these

    4.   1   2   3   4   5

5. Four of Mr. Yokimo's brothers live in America; two more brothers live in Japan. How many brothers does Mr. Yokimo have in all?

    (1) 2   **(2)** 7   **(3)** 5   **(4)** 8
    (5) none of these

    5.   1   2   3   4   5

6. The parks department repaired 32 picnic tables on Tuesday, 26 on Wednesday, and 20 more on Thursday. How many tables did the department repair altogether?

    (1) 87   **(2)** 79   **(3)** 80   **(4)** 78
    (5) none of these

    6.   1   2   3   4   5

7. You bought 3 packs of gum, 2 magazines, a candy bar, and 3 oranges to take on a train ride. How many items did you buy in all?

    (1) 7   **(2)** 5   **(3)** 8   **(4)** 9
    (5) none of these

    7.   1   2   3   4   5

8. The rent is $248 a month. Food for the family is $250 a month. What is the total cost of these two expenses?

    (1) $450   **(2)** $498   **(3)** $500   **(4)** $488
    (5) none of these

    8.   1   2   3   4   5

9. There were 42 tomato plants, 13 heads of lettuce, and 22 strawberry plants in a garden. Ten more tomato plants were added later. How many plants are there in the garden altogether?

9. 1 2 3 4 5

(1) 77   (2) 87   (3) 52   (4) 45
(5) none of these

10. At the rally, 342 people voted for the labor agreement. The number of people who voted against it was 427. How many votes were there in all?

10. 1 2 3 4 5

(1) 769   (2) 186   (3) 768   (4) 679
(5) none of these

Check your answers. If all your answers are right, begin Chapter 3. If one or two answers are wrong, make sure that you understand why the answers given are the right ones. Then go to Chapter 3. If you made more than two mistakes, you should read Chapter 2 again. Then start Chapter 3.

ANSWERS

1.1   8 hangers in all

2.1. 9

3.1. (a) 4   (b) 3   (c) 8   (d) 7   (e) 2 + 3 = 5   (f) 8   (g) 11   (h) 12
              +2
             ───
               6

4.2. 87; 37

4.3 (a)   78   (b) 87   (c) 89   (d) 38   (e) 57
            + 21
            ────
              99

4.5. (a)   34   (b)   26   (c)   57
             + 52   + 31   + 40
             ────   ────   ────
               86   57   97

5.1. 665

5.2. (a)   456   (b) 589   (c) 847   (d) 279   (e) 768   (f) 897   (g) 759   (h) 288
             + 123
             ─────
               579

6.1. 8

6.2. **(a)** 3   **(b)** 8   **(c)** 9   **(d)** 8   **(e)** 7
            2
         + 4
          ——
          9

6.3. **(a)** 2   **(b)** 9   **(c)** 7   **(d)** 9   **(e)** 9
            1
            1
         + 4
          ——
          8

7.1. **(a)** 22   **(b)** 88   **(c)** 86   **(d)** 79   **(e)** 89
         35
       + 12
        ——
        69

8. **(a)** 18   **(b)** 77   **(c)** 978   **(d)** 87   **(e)** 13

9. **(a)** Altogether: 3   There were 8 touchdowns.
                 0
                 2
              + 3
             ——
               8

(b) Increased: $3   The price of coffee this month is $4.
             + $1
           ——
           $4

(c) Total: 12   The total number of items purchased was 18.
            4
         + 2
        ——
        18

(d) In all: 12   There were 44 boxes in all.
        + 32
        ——
        44

(e) Gained: 43   The dog now weighs 55 pounds.
        + 12
        ——
        55

(f) More: 24   There are 37 people now on the payroll.
        + 13
        ——
        37

(g) Sum: 42
            13
         + 24
         ——
            79

The sum is 79.

1. (3)   327
       + 562
       ——
         889

2. (2)   12
       + 4
       ——
          16

3. (3) $45
       + $33
       ——
         $78

4. (4)   2
          4
          1
       + 2
       ——
          9

5. (5)   4
       + 2
       ——
          6

6. (4)   32
          26
       + 20
       ——
          78

7. (4)   3
          2
          1
       + 3
       ——
          9

8. (2) $248
       + $250
       ——
         $498

9. (2)   42
          13
          22
       + 10
       ——
          87

10. (1)   342
        + 427
        ——
          769

THREE

Addition: Part II

1. CARRYING OVER

A nurse needs 9 pillows for one hospital room and 8 pillows for another room. How many pillows in all does the nurse need?

$$\text{Add:} \quad \begin{array}{r} 9 \\ + 8 \\ \hline \end{array}$$

The 1 is CARRIED OVER to the tens column because 9 + 8 adds up to a two-digit number: 17
The nurse needs 17 pillows.

tens	units
	9
+	8
①	7

1.1. Test Yourself

Do this addition:

$$\begin{array}{r} 8 \\ + 5 \\ \hline \end{array}$$

Check your answer. If it is right, go to section 1.2. If it is wrong, go to section 1.1A.

1.1A. From the addition chart, you know that 8 + 5 = 13. Since 13 is a two-digit number, you need a new column (the tens column, on the left side of the units column) to write the addition in column form:

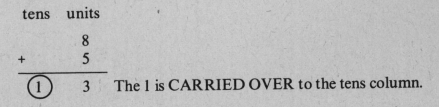

tens	units
	8
+	5
①	3

The 1 is CARRIED OVER to the tens column.

1.2. Practice

(a) 9   (b) 4   (c) 6   (d) 7   (e) 5
    + 7   + 8   + 6   + 8   + 5
    ___   ___   ___   ___   ___

Check your answers and go to section 2.

2. CARRYING OVER WITH TWO-DIGIT NUMBERS

If the numbers in a column add up to 10 or more, you must CARRY OVER to the next column.

First example: carrying over from the units column to the tens column:

Add the units column: 5
                        + 7
                        ____
                         12

tens	units
2 .	5
+ 3	7

Carry the 1 from the 12 over to the top of the tens column.

Add the 1 with the other numbers in the tens column: ①
                                                       2
                                                     + 3
                                                     ___
                                                       6

①	
2	5
+ 3	7
	2

Your answer is 62.

①	
2	5
+ 3	7
6	2

Second example: carrying over from the tens column to the hundreds column.

Add the units column: 5 + 4 = 9

Add the tens column: 5 + 8 = 13

Carry the 1 from the 13 over to the hundreds column.

Your answer is 139.

hundreds	tens	units
	5	5
+	8	4

hundreds	tens	units
	5	5
+	8	4
		9

hundreds	tens	units
	5	5
+	8	4
①	3	9

2.1. Test Yourself

Do these two additions:

$$\begin{array}{r} 25 \\ + 49 \\ \hline \end{array} \qquad \text{and} \qquad \begin{array}{r} 42 \\ + 81 \\ \hline \end{array}$$

Check your answers. If both are right, go to section 2.2. If one is wrong, go to section 2.1A.

2.1A. For the first addition, you carry the 1 over to the top of the tens column.

$$\begin{array}{cc} \text{tens} & \text{units} \\ \textcircled{1} & \\ 2 & 5 \\ + 4 & 9 \\ \hline 7 & 4 \end{array}$$

Then you add all the numbers in the tens column.

For the second addition, you need a new column on the left side of the tens column: the hundreds column. You carry the 1 from the tens column to the hundreds column:

$$\begin{array}{ccc} \text{hundreds} & \text{tens} & \text{units} \\ & 4 & 2 \\ + & 8 & 1 \\ \hline \textcircled{1} & 2 & 3 \end{array}$$

2.2. Practice

DON'TCROWDYOURWORK!   Be sure to leave space for a new column on the left side of additions. You never know when you will need to carry over to a new column.

(a)	(b)	(c)	(d)	(e)
$\begin{array}{r} 35 \\ + 94 \\ \hline \end{array}$	$\begin{array}{r} 20 \\ + 89 \\ \hline \end{array}$	$\begin{array}{r} 63 \\ + 52 \\ \hline \end{array}$	$\begin{array}{r} 29 \\ + 90 \\ \hline \end{array}$	$\begin{array}{r} 82 \\ + 45 \\ \hline \end{array}$

Check your answers. If they are all correct, go to section 3. If one or more are wrong, read section 2 again.

3. CARRYING OVER WITH THREE-DIGIT NUMBERS

The way to do this is exactly the same as for two-digit numbers.

First example: carrying over from the units column to the tens column.

Add the units column:

$$\begin{array}{r} 5 \\ + 7 \\ \hline 12 \end{array}$$

hundreds	tens	units
2	3	5
+ 3	1	7

	①	
2	3	5
+ 3	1	7
		2

Carry the 1 from the 12 over to the top of the tens column.

Add the tens column (don't forget the 1 you carried over!).

Add the hundreds column.

Your answer is 552.

	①	
2	3	5
+ 3	1	7
5	5	2

Second example: carrying over from the tens column to the hundreds column.

H	T	U
5	7	4
+ 2	8	2
		6

Add the units column.

Add the tens column.

Carry the 1 over to the top of the hundreds column.

Add the hundreds column.

Your answer is 856.

①		
5	7	4
+ 2	8	2
	5	6

①		
5	7	4
+ 2	8	2
8	5	6

Third example: carrying over from the hundreds column to the thousands column.

Add the units column. Add the tens column.

Add the hundreds column:

$$\begin{array}{r} 5 \\ + 9 \\ \hline 14 \end{array}$$

TH	H	T	U
	5	5	3
+	9	3	2
		8	5

Carry the 1 from the 14 over to the thousands column.

Your answer is 1,485.

TH	H	T	U
	5	5	3
+	9	3	2
①	4	8	5

3.1. Test Yourself

Remember to leave some space on the left side of your additions, in case you need a column for the thousands.

     (a)   356   (b)   291   (c)   845
          + 219   + 357   + 613

Check your answers. If they are all correct, go to section 3.2. If not, read section 3 again.

3.2. Practice

     (a)   613   (b)   267   (c)   523   (d)   492   (e)   639
          + 745   + 248   + 248   + 113   + 245

Check your answers. Then go on to section 3.3.

3.3. Practice

You can carry over with any number. Do these additions with four-digit numbers:

     (a)   6,542   (b)   3,227   (c)   4,359   (d)   8,634   (e)   7,464
        + 5,226   + 2,930   + 1,319   + 5,251   + 1,281

Check your answers.

4. CARRYING OVER MORE THAN ONCE

Carrying over can happen in more than one column.

Example: 6,986 + 458
Add the units column: 6
                    + 8
                      14

TH	H	T	U
			①
6,	9	8	6
+	4	5	8
			4

Carry the 1 over to the top of the tens column.

Add the tens column: 1
                               8
                          + 5
                          ――
                         14

Carry the 1 over to the top of the hundreds column.

Add the hundreds column: 1
                               9
                          + 4
                          ――
                         14

Carry the 1 over to the thousands column.

Add the thousands column: 1
                               + 6
                          ――
                         7

Your answer is 7,444.

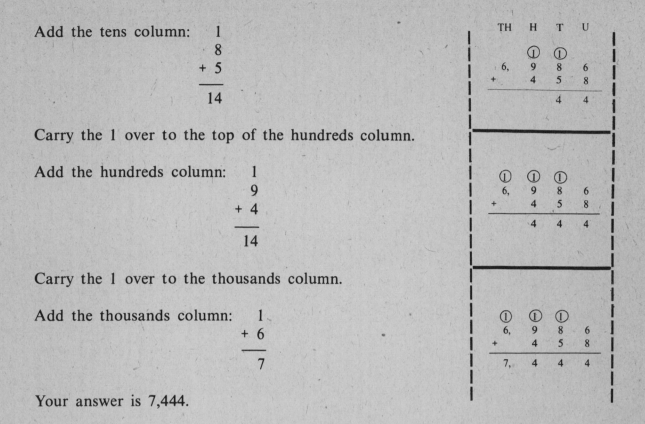

5. CARRYING OVER A NUMBER OTHER THAN 1

If you are adding more than two numbers, it may happen that the number you have to carry over is not 1, but 2, or even a larger number. Here is an example.

Add the two first numbers: 7 + 8 = 15

Add the total to the last number: 15 + 6 = 21

Now, in order to write your answer, you must carry the 2 to the tens column.

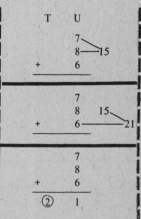

Your answer is 21.

6. YOU CAN ADD ANY NUMBERS

In Chapters 2 and 3, you have been learning how to add ANY NUMBERS. No matter how many digits they have, no matter how many of them you need to add together, all you have to do is remember the rules you learned so far.

7. ADDITION REMINDERS

Be sure to put the numbers you are adding in the correct columns before you add.

234 + 56 + 2,643 + 543   To add, write in column form:

```
      234
       56
    2,643
  +   543
  -------
```

Put units under units, tens under tens, hundreds under hundreds, thousands under thousands.

Leave some space on the left side of your addition in case you need a new column to carry over.

Commas are put after every three digits, starting from the right.

8. MORE PRACTICE

(a)   16
      + 29

(b)   258
      + 481

(c)   426
        395
      + 562

(d)   5,624
    + 2,164

(e) 97 + 36 + 48 =

(f)   17
        26
        43
      + 35

(g)   824
      + 351

(h)   987
        645
      + 294

(i)   536
         49
      + 821

(j) 111 + 243 + 14 =

9. WORD PROBLEMS

Before you start these problems, look at Chapter 2 again. That will remind you of the ADDITION WORD CLUES. Reread HOW TO SOLVE WORD PROBLEMS to remind you of the steps to follow.

1. If 29 cars need no-lead gasoline and 180 need regular gasoline, how many cars in all need gas?

   (1) 289   **(2)** 92   **(3)** 282   **(4)** 209
   (5) none of these

2. The O'Briens bought a washing machine in 1954. 16 years later they bought a new one. What year did they buy a new machine?
   (1) 1960   **(2)** 1970   **(3)** 1971   **(4)** 1969
   (5) none of these

3. The township was selling three parcels of land: one parcel was 840 acres, one parcel was 1,235 acres and another was 1,150 acres. How many acres in all was the township selling?
   (1) 3,825   **(2)** 4,225   **(3)** 3,225   **(4)** 3,525
   (5) none of these

4. A sweepstakes offers 1 grand prize, 77 second prizes and 1,977 third prizes. What is the total number of prizes to be awarded?
   (1) 2,050   **(2)** 3,855   **(3)** 2,055   **(4)** 2,578
   (5) none of these

5. Ford Motor Company recalled 2,400 Pintos, 3,700 Mustangs and 5,849 Fairlanes. How many cars were recalled in all?
   (1) 1,949   **(2)** 13,949   **(3)** 12,999   **(4)** 11,599
   (5) none of these

6. A branch of the city's library had 8,960 books for adults, 2,450 books for teenagers, and 1,580 books for children. How many books were there in all?
   (1) 12,900   **(2)** 12,880   **(3)** 12,990   **(4)** 13,990
   (5) none of these

7. The sanitation department issued 2,800 summonses for unleashed dogs and 8,000 tickets for littering. How many tickets in all were given out?
   (1) 10,080   **(2)** 11,080   **(3)** 28,800   **(4)** 10,800
   (5) none of these

8. Horse of the Year, Forego, earned $491,701 in 1976, $286,740 in 1977, and $15,000 in 1978. How much did Forego earn in the three years?
   (1) $793,441   **(2)** $930,441   **(3)** $753,441   **(4)** $893,440
   (5) none of these

1.   1   2   3   4   5
2.   1   2   3   4   5
3.   1   2   3   4   5
4.   1   2   3   4   5
5.   1   2   3   4   5
6.   1   2   3   4   5
7.   1   2   3   4   5
8.   1   2   3   4   5

Check your answers. If all your answers are right, begin Chapter 4. If one or two answers are wrong, make sure that you understand why the answers given are the right ones and then go on to Chapter 4. If you made three mistakes or more, you should read Chapter 3 again. Then start Chapter 4.

10. ANSWERS

1.1. 13

1.2. (a) 16   (b) 12   (c) 12   (d) 15   (e) 10

2.1. 74; 123

2.2. (a) 129   (b) 109   (c) 115   (d) 119   (e) 127

3.1. (a) 575   (b) 648   (c) 1,458

3.2. (a) 1,358   (b) 515   (c) 771   (d) 605   (e) 884

3.3. (a) 11,768   (b) 6,157   (c) 5,678   (d) 13,885   (e) 8,745

8. (a) 45   (b) 739   (c) 1,383   (d) 7,788   (e) 181

    (f) 121   (g) 1,175   (h) 1,926   (i) 1,406   (j) 368

9. Word Problems

1. (4)
$$
\begin{array}{r}
29 \\
+\ 180 \\
\hline
209
\end{array}
$$

2. (2)
$$
\begin{array}{r}
1954 \\
+\ 16 \\
\hline
1970
\end{array}
$$

3. (3)
$$
\begin{array}{r}
840 \\
1,235 \\
+\ 1,150 \\
\hline
3,225
\end{array}
$$

4. (3)
$$
\begin{array}{r}
1 \\
77 \\
+\ 1,977 \\
\hline
2,055
\end{array}
$$

5. (5)
$$
\begin{array}{r}
2,400 \\
3,700 \\
+\ 5,849 \\
\hline
11,949
\end{array}
$$

6. (3)
$$
\begin{array}{r}
8,960 \\
2,450 \\
+\ 1,580 \\
\hline
12,990
\end{array}
$$

7. (4)
$$
\begin{array}{r}
2,800 \\
+\ 8,000 \\
\hline
10,800
\end{array}
$$

8. (1)
$$
\begin{array}{r}
\$491,701 \\
286,740 \\
+\ 15,000 \\
\hline
\$793,441
\end{array}
$$

FOUR

Subtraction: Part I

1. WHAT IS SUBTRACTION?

SUBTRACTION is taking away one amount from another. When you subtract, your answer is less than the amount you had in the beginning. You had 7 books. You gave away 4 of them. How many do you have left?

   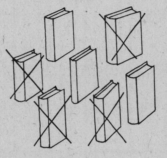   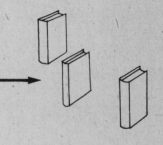

7 books   4 given away   3 are left

There are two ways of writing this subtraction.

You can write it either this way: 7 – 4 = 3

or this way:
$$\begin{array}{r} 7 \\ -\ 4 \\ \hline 3 \end{array}$$

To read this subtraction you say: 7 minus 4 is 3.

1.1. Test Yourself

There are 8 tables in a coffee shop. There are people sitting at 2 of them. How many tables are still free?

Check your answer. If it is right, go to section 1.2. If it is wrong or if you don't know what to do, go to section 1.1A.

1.1A. Subtracting means taking away. If you take 2 tables away from 8 tables, 6 tables are left:

So the answer is 6.

1.2. The two ways of writing your subtraction are:

$$8 - 2 = 6 \quad \text{or} \quad \begin{array}{r} 8 \\ - 2 \\ \hline 6 \end{array}$$

2. SUBTRACTION IS RELATED TO ADDITION

There are 5 chairs in the dining room. You take 2 of them to the bedroom. How many chairs are left in the dining room?

$$\begin{array}{r} 5 \\ - 2 \\ \hline 3 \end{array}$$

Then you bring back the 2 chairs from the bedroom to the dining room. How many chairs are there now in the dining room?

$$\begin{array}{r} 3 \\ + 2 \\ \hline 5 \end{array}$$

3. HELPFUL HINTS FOR SUBTRACTION

The MINUS SIGN is used for subtraction. It tells you to SUBTRACT one number away from another. It looks like this: −

When you subtract, first write the number you subtract from, then the number you take away:

9 – 5 = 4   or   9
                  – 5
              ——

           4   The 5 is being subtracted from 9

                  Your answer is 4.

The answer in subtraction is called the DIFFERENCE. To check your subtraction, here is all you do:

```
  9
– 5     subtract
——
  4

  4     add the answer
+ 5     and the number
——      you took away.
  9
```

Do they add up to the number subtracted from? You got it right.

3.1. Practice

Subtract and check

    (a)   7   (b)   6   (c)   8   (d)   9   (e)   5
         –5   –4   –0   –2   –4

Check your answers. If they are all right, go to section 4. If one is wrong, or if you don't know what to do, go to section 3.1A.

3.1A. You may have been confused by the zeros: 8
                                          – 0

Taking away zero means taking away nothing at all: if you have 8 books, and you give 0 books away, it means that you don't give anything away and you keep your 8 books. So   8
     –0
     ——
     8

If you have 5 paper clips, and you give 0 paper clips away, it means that you keep all your paper clips: so 5 – 0 = 5.

Look at this subtraction: 6 – 6 =

If you have 6 erasers and you give 6 erasers away, you are left with nothing at all. So 6 – 6 = 0.

For the other subtractions, you may find it helpful to make drawings. Here is a drawing you can make for

$$\begin{array}{r} 7 \\ -\ 5 \\ \hline \end{array}$$

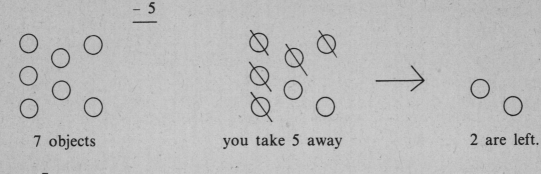

7 objects   you take 5 away   2 are left.

So

$$\begin{array}{r} 7 \\ -\ 5 \\ \hline 2 \end{array}$$

Try to remember these SUBTRACTION FACTS so that later on you can subtract more quickly.

4. SUBTRACTING TWO–DIGIT NUMBERS

On Tuesday 68 tickets to the show were sold. If 45 tickets were sold in the morning, how many were sold in the afternoon?

To subtract a two-digit number from another two-digit number, use the column form. Put units under units, and tens under tens.

Subtract the units column:
$$\begin{array}{r} 8 \\ -\ 5 \\ \hline 3 \end{array}$$

Subtract the tens column:
$$\begin{array}{r} 6 \\ -\ 4 \\ \hline 2 \end{array}$$

Your answer is 23.

tens column	units column
6	8
– 4	5
	3
6	8
– 4	5
2	3
2	3
+ 4	5
6	8

There were 23 tickets sold in the afternoon. Check by adding your answer and the number you took away.

4.1. Test Yourself

Do this subtraction: 59   and check it by adding.
                 − 12

Check your answer. If it is right, go to section 4.2. If it is wrong or if you don't know what to do, go to section 4.1A.

4.1A. Before you start the subtraction, be sure to separate the numbers into two columns: the units column and the tens column. First subtract the units column, then subtract the tens column

tens	units			tens	units
5	9	Your answer is 47.		4	7
− 1	2	You check it by adding:		+ 1	2
4	7			5	9

Go to section 4.2.

4.2. Sometimes you don't need to subtract in the tens column. Here is an example:

    49
   − 3

Subtract the units column: 9
                            − 3
                            6

tens	units
4	9
−	3
	6
4	9
−	3
4	6
4	6
+	3
4	9

In the tens column, there is nothing to subtract from 4: just bring it down.

Your answer is 46.

Check.

Sometimes you get a zero in the tens column so the answer is a one-digit number.

Here is an example:

Subtract the units column.
Subtract the tens column.

Drop the zero.
Your answer is 5.

tens	units
2	7
− 2	2
2	7
− 2	2
0	5
2	7
− 2	2
	5

4.3. Practice

Subtract and check

(a)	63	**(b)**	49	**(c)**	27	**(d)** 27 − 15 =
	− 21		− 15		− 3	**(e)** 14 − 3 =

Check your answers.

5. SUBTRACTING THREE OR MORE DIGITS

A window washer cleaned 124 windows in a building that had 256 windows. How many windows were still dirty?

$$256 - 124$$

Put your subtraction in the column form: units under units, tens under tens, hundreds under hundreds.

Subtract the units column.

Subtract the tens column.

Subtract the hundreds column.

Your answer is 132.
There are 132 windows that are still dirty.
Check.

H	T	U
2	5	6
− 1	2	4
		2

2	5	6
− 1	2	4
	3	2

2	5	6
− 1	2	4
1	3	2

1	3	2
+ 1	2	4
2	5	6

5.1. Test Yourself

Subtract and check:

(a)	375	**(b)**	6,493
	− 134		− 3,240

Check your answers. If they are both right, go to section 5.1B. If one is wrong or if you don't know what to do, go to section 5.1A.

5.1A. Rules for subtraction:

Put in column form: 857 − 342
$$\begin{array}{r} 857 \\ -342 \\ \hline \end{array}$$

Place units under units, tens under tens, hundreds under hundreds.

$$\begin{array}{r} 857 \\ -342 \\ \hline \end{array}\qquad \begin{array}{r} 857 \\ -342 \\ \hline \end{array}$$

This way   NOT this way

Start subtracting units, then tens, then hundreds.

$$\begin{array}{r} 857 \\ -342 \\ \hline 515 \end{array}$$

Check.

$$\begin{array}{r} 342 \\ +515 \\ \hline 857 \end{array}$$

Correct your work and go to section 5.1B.

5.1B. Sometimes you do not need to subtract in all the columns. For example:

3,269 − 48

Put your subtraction in the column form.

Subtract the units column, then the tens column.

In the hundreds column, there is nothing to subtract from 2: just write it down.

In the thousands column, there is nothing to subtract from 3: just write it down.
Check.

TH	H	T	U
3,	2	6	9
−		4	8
		2	1
3,	2	6	9
−		4	8
3,	2	4	8
3,	2	4	8
+		2	1
3,	2	6	9

5.2. Practice

Subtract and check:

(a) $\begin{array}{r} 653 \\ -412 \\ \hline \end{array}$ (b) $\begin{array}{r} 937 \\ -15 \\ \hline \end{array}$ (c) $\begin{array}{r} 7,839 \\ -5,624 \\ \hline \end{array}$ (d) 649 − 237 =

(e) 4,385 − 203 =

Check your answers. If they are all right, go to section 6. If not, read section 5 again, and then go to section 6.

6. ZERO IS IMPORTANT

Remember that zero is a number. When you subtract and your answer is zero, be sure to put it in place. Zero is as important as any other number. You have $68 and you spend $48. Subtract to find out what you have left.

$$68 - 48 =$$

Put your subtraction in the column form.

Subtract units.

Subtract tens.

Your answer is 20.
You have $20 left.

Check.

	T	U
	$6	8
–	4	8
		0

	T	U
	$6	8
–	4	8
	$2	0

	T	U
	$2	0
+	4	8
	$6	8

Zero makes a big difference:

correct subtraction:

$$\begin{array}{r} \$68 \\ -\ 48 \\ \hline \$20 \end{array}$$

incorrect subtraction:

$$\begin{array}{r} \$68 \\ -\ 48 \\ \hline \$2 \end{array}$$

Which amount would you rather have? Always put zero in its place.

6.1. Practice

Subtract and check:

(a) 73
  – 43

(b) 257
  – 147

(c) 8,965
  – 2,165

(d) 5,643
  – 4,621

(e) 48
  – 28

Check your answers.

6.2. More Practice

Subtract and check. When a subtraction is not in the column form, put it in the column form first.

(a) 6
  – 2

(b) 38
  – 12

(c) 7,893
  – 5,462

(d) 87
  – 41

(e) 98
  – 48

(f) 264
   − 163

(g) 908
   − 206

(h) 81
   − 80

(i) 52
   − 12

(j) 47 − 26 =

Check your answers.

7. WORD PROBLEMS—SUBTRACTION

SUBTRACTING is taking one amount away from another. Subtraction word problems are easier to solve if you look for subtraction word clues. Some subtraction word clues are:

(a) DIFFERENCE—Soap powder at the corner Quick-Stop costs 98 cents. At the supermarket the same brand costs 87 cents. What is the difference in price?

(b) REMAINS—A rancher sold 356 cattle from his stock of 457. How much of the stock remains on the ranch?

(c) DECREASE—Eggs dropped from 89 cents to 69 cents a dozen. How much did the price of eggs decrease?

(d) REDUCED—There were 6,758 firefighters before lay-offs, and 4,648 after lay-offs. The number of firefighters was reduced by how many?

(e) FEWER—Last year 239 workers went to the training program but this year only 217 went. How many fewer workers went this year?

(f) LESS—E-Z Drive Auto School offers driving lessons for 60 dollars. Safety First Auto School offers lessons for 97 dollars. How much less are lessons at E-Z Drive?

(g) MORE—The report said that 5,645 people listen to the news on the radio, whereas 9,798 watch the news on TV. How many more people watch TV than listen to the radio?

DIFFERENCE—   98
            − 87
            ____
             11   The difference is 11 cents.

REMAINS—   457
         − 356
         ____
          101   101 cattle remain on the ranch.

DECREASE—   89
          − 69
          ____
           20   The price of eggs decreased 20 cents.

REDUCED—   6,758
         − 4,648
         ____
          2,110   The number of firefighters was reduced by 2,110.

```
FEWER—     239
          - 217
          ————
            22   22 fewer workers went this year.

LESS—      $97
          - 60
          ————
          $37   The lessons are $37 less at E–Z Drive.

MORE—    9,798
       - 5,645
       ———————
         4,153   4,153 more people watch TV than listen to the radio.
```

7.1. WORD PROBLEMS

Before you begin these problems, look at HOW TO SOLVE WORD PROBLEMS in Chapter 2.

1. A 50-foot roll of snow fence cost $11 three years ago. The same snow fence now costs $24. How much has the price increased?
(1) $3   (2) $10   (3) $13   (4) $15
(5) none of these

1.   1   2   3   4   5

2. Singer Sewing Company celebrated its 126 years in business in 1978. What was the first year of business for the Singer Sewing Machine Company?
(1) 1826   (2) 1854   (3) 1852   (4) 1850
(5) none of these

2.   1   2   3   4   5

3. The recipe says to bake the meat at 375 degrees for 2 hours, then turn down the oven to 250 degrees for another hour. How many degrees is the oven temperature reduced?
(1) 125   (2) 120   (3) 25   (4) 175
(5) none of these

3.   1   2   3   4   5

4. A three-piece set of living room furniture was reduced from $299 to $248. The price was reduced by how much?
(1) $48   (2) $51   (3) $57   (4) $37
(5) none of these

4.   1   2   3   4   5

5. Sophie Balgeria started dieting when she weighed 175 pounds. When she got down to 130 pounds she stopped dieting. How many pounds did Sophie lose?
(1) 40   (2) 35   (3) 54   (4) 45
(5) none of these

5.   1   2   3   4   5

6. A resort town collected $189,000 by charging $1.00 a
person to enter the beach. Of this money, $126,000 was
spent on repairing beach fences. How much money was
left over for other expenses?
(1) $63,000   (2) $75,000   (3) $636.00   (4) $36.00
(5) none of these

6. 1   2   3   4   5

7. After four years, a summer lifeguard reaches maximum
salary of $19 a day. The beginning pay is $16 a day. How
much does a lifeguard's salary increase over four years?
(1) $35   (2) $154   (3) $23   (4) $3
(5) none of these

7. 1   2   3   4   5

8. A group of 5,647 people attended Northville's Baby Parade.
Of those people, 1,235 were under 3 years old. How many
viewers were over 3 years old?
(1) 4,412   (2) 4,192   (3) 6,412   (4) 4,442
(5) none of these

8. 1   2   3   4   5

Check your answers. If all your answers are right, begin Chapter 5. If one or two
problems are wrong, make sure that you understand the right answers, then go to
Chapter 5. If you made more than two mistakes, you should read Chapter 4 again.
Then start Chapter 5.

8. ANSWERS

1.1. 6

3.1. (a) 2   (b) 2   (c) 8   (d) 7   (e) 1

4.1. 47

4.3. (a) 42   (b) 34   (c) 24   (d) 12   (e) 11

5.1. (a) 241   (b) 3,253

5.2. (a) 241   (b) 922   (c) 2,215   (d) 412   (e) 4,182

6.1. (a) 30   (b) 110   (c) 6,800   (d) 1,022   (e) 20

6.2. (a) 4   (b) 26   (c) 2,431   (d) 46   (e) 50

     (f) 101   (g) 702   (h) 1   (i) 40   (j) 21

7.1. Word Problems

1. (3)　　$24
　　　　　－　11
　　　　　―――――
　　　　　　$13

2. (3)　　1978
　　　　　－　126
　　　　　―――――
　　　　　　1852

3. (1)　　375
　　　　　－　250
　　　　　―――――
　　　　　　125 degrees

4. (2)　　$299
　　　　　－　248
　　　　　―――――
　　　　　　$ 51

5. (4)　　175
　　　　　－　130
　　　　　―――――
　　　　　　45

6. (1)　　189,000
　　　　　－　126,000
　　　　　―――――
　　　　　　63,000

7. (4)　　$19
　　　　　－　16
　　　　　―――――
　　　　　　$ 3

8. (1)　　5,647
　　　　　－　1,235
　　　　　―――――
　　　　　　4,412

FIVE

Subtraction: Part II

1. BORROWING

You have 17 pages to read in a book. You have already read 9 pages. How many pages do you still have to read?

    17 – 9

Put your subtraction in the column form.
Subtract the units column. . . . but 9 units are more than 7 units.

BORROW a ten and put it in the units column. Since one ten is ten units, you now have seventeen units.
Now you can subtract: 17 – 9 = 8.

There is nothing left in the tens column. Your answer is 8.
You still have eight pages to read.

Check.

T	U
1	7
–	9

T	U
1̸	17 / 7̸
–	9
	8

	U
	8
+	9
1	7

1.1. Test Yourself

Subtract and check: 15 – 8 =

Check your answer. If it is right, go to section 1.2. If it is wrong or if you don't know what to do, go to section 1.1A.

1.1A. First put your subtraction in the column form:

      15
    – 8
    ———

You cannot subtract the units column because 8 is more than 5. So borrow a ten and put it in the units column. You have 15 units. Now you can subtract 8:

$$\begin{array}{r} \overset{15}{\cancel{1}\,\cancel{5}} \\ -\ \ 8 \\ \hline 7 \end{array}$$

Your answer is 7. Check it by adding.
Go to section 1.2.

1.2. Practice

     (a) $\begin{array}{r} 17 \\ -\ 8 \\ \hline \end{array}$ **(b)** $\begin{array}{r} 15 \\ -\ 7 \\ \hline \end{array}$ **(c)** $\begin{array}{r} 18 \\ -\ 9 \\ \hline \end{array}$ **(d)** $12 - 9 =$
                                                (e) $15 - 6 =$

Check your answers.

1.2A. Sometimes after you have borrowed, there are still tens in the tens column. A bus carrying 54 passengers stops and 6 people get off. How many people remain on the bus? $54 - 6$

Put your subtraction in the column form.
Subtract the units column . . . but 6 is more than 4.

Borrow a ten from the tens column and put it in the units column. A ten is ten units, so you now have 14 units.
Subtract the units column: $14 - 6 = 8$.

4 is left in the tens column.
There is nothing to take away from 4, so just bring it down.

Your answer is 48.
There are 48 people remaining on the bus.

Check it by adding.

$$\begin{array}{c|cc} & T & U \\ & 5 & 4 \\ - & & 6 \\ \hline \end{array}$$

$$\begin{array}{c|cc} & \overset{4}{\cancel{5}} & \overset{14}{\cancel{4}} \\ - & & 6 \\ \hline & & 8 \end{array}$$

$$\begin{array}{c|cc} & \overset{4}{\cancel{5}} & \overset{14}{\cancel{4}} \\ - & & 6 \\ \hline & 4 & 8 \end{array}$$

$$\begin{array}{c|cc} & \overset{①}{4} & 8 \\ + & & 6 \\ \hline & 5 & 4 \end{array}$$

2. MORE BORROWING FROM A TWO-DIGIT NUMBER

74 − 35 =

Put your subtraction in the column form.
Subtract the units column . . . but 5 is more than 4, so you
can't.

Borrow a ten from the tens column and put it in the units
column. You now have 14 units.
Subtract the units column: 14 − 5 = 9.

Subtract the tens column: 6 − 3 = 3

Your answer is 39.

Check by adding.

T	U
7	4
− 3	5

6̸7	1̸4 4
− 3	5
	9

6̸7	1̸4 4
− 3	5
3	9

①3	9
+ 3	5
7	4

When you borrow, cross the numbers out but do not erase them. It is better to show all
the changes you make. It makes the subtracting and the checking easier.

2.1. Test Yourself

Subtract and check:

```
    34          65
  −  8        − 29
  ____        ____
```

Check your answers. If they are both right, go to section 2.2. If one of them is wrong,
read sections 1.2A and 2 again. Then go to section 2.2.

2.2. Practice

Subtract and check. Some problems have been started.

```
        5 14           6 15          8 17          7 11
(a)     6̸ 4̸    (b)     7̸ 5̸   (c)    9̸ 7̸   (d)     8̸ 1̸     (e) 61 − 32 =
      − 2 6          − 2 9         − 5 8          − 3 3      (f) 24 − 15 =
      _____        _____        _____        _____      (g) 73 − 57 =
                                                             (h) 42 − 38 =
```

Check your answers.

3. BORROWING FROM A THREE-DIGIT NUMBER

You may have to borrow from the hundreds column. Here is an example.

638 – 355

Put your subtraction in the column form.
Subtract the units column: 8 – 5 = 3.

Subtract the tens column . . . but 5 is more than 3.
Borrow one hundred from the hundreds column and put it in
the tens column. You now have 13 tens in the tens column.
Subtract the tens column: 13 – 5 = 8.

Subtract the hundreds column: 5 – 3 = 2.

Your answer is 283.

Check.

	H	T	U
	6	3	8
–	3	5	5
			3

	H	T	U
	5̶6̶	13̶3̶	8
–	3	5	5
		8	3

	5̶6̶	13̶3̶	8
–	3	5	5
	2	8	3

	①2	8	3
+	3	5	5
	6	3	8

3.1. Practice

(a)
```
  4 13
  5̶ 3̶ 4
- 2 8 0
```

(b)
```
  6 13
  7̶ 3̶ 8
- 4 5 2
```

(c)
```
  3 12
  4̶ 2̶ 5
- 1 7 4
```

(d) 627 – 253 =

(e) 727 – 385 =

Check your answers.

4. BORROWING FROM A FOUR-DIGIT NUMBER

You may have to borrow from the thousands column.
5,376 – 3,853

Put in column form.
Subtract units column.
Subtract tens column.
Subtract hundreds column . . . but 8 is more than 3, so you
can't do it.

	TH	H	T	U
	5,	3	7	6
–	3,	8	5	3
			2	3

Borrow one thousand from the thousands column, and put it in the hundreds column: you now have 13 hundreds.
Subtract hundreds column: $13 - 8 = 5$.

Subtract thousands column: $4 - 3 = 1$.

Your answer is 1,523.

Check.

	4 ̶5̶	13 ̶3̶	7	6
-	3,	8	5	3
		5	2	3

	4 ̶5̶	13 ̶3̶	7	6
-	3,	8	5	3
	1,	5	2	3

	①1,	5	2	3
+	3,	8	5	3
	5,	3	7	6

4.1. Practice

(a)
$$\begin{array}{r} {}^{7}\!\!\!\not{8},\ {}^{13}\!\!\!\not{3}\ 7\ 2 \\ -\ 2,5\ 4\ 0 \\ \hline \end{array}$$

(b)
$$\begin{array}{r} 4,6\ 3\ 5 \\ -\ 1,8\ 2\ 2 \\ \hline \end{array}$$

(c) $5,270 - 1,940 =$
(d) $6,342 - 2,831 =$
(e) $7,489 - 4,876 =$

Check your answers.

4.1A. BORROWING MORE THAN ONCE
You may have to borrow more than once in the same subtraction problem.
$$8,347 - 5,989$$

Put in column form.
Subtract units . . . but 9 is more than 7.
Borrow a ten from the tens column and put it in the units column. You now have 17 units.
Subtract units: $17 - 9 = 8$.

TH	H	T	U
		3 ̶4̶	17 ̶7̶
8,	3	4	7
- 5,	9	8	9
			8

Subtract tens . . . but 8 is more than 3.
Borrow a hundred from the hundreds column and put it in the tens column. You now have 13 tens.
Subtract tens: $13 - 8 = 5$.

	2 ̶3̶	13 ̶4̶	17
8,	3	4	7
- 5,	9	8	9
		5	8

Subtract hundreds . . . but 9 is more than 2.
Borrow a thousand from the thousands column and put it in the hundreds column. You now have 12 hundreds.
Subtract hundreds: $12 - 9 = 3$.
Subtract thousands: $7 - 5 = 2$.
Your answer is 2,358.

	7 ̶8̶	12 ̶2̶	13	17
- 5,	9	8	9
2,	3	5	8

	①2,	①3,	①5	8
+ 5,	9	8	9
8,	3	4	7

Check.

When you keep your work well-spaced and neat, it is much easier to see exactly what is going on in a problem. Leave some space on top of your subtraction in case you need to borrow several times.

4.2. Test Yourself

Finish this subtraction and check it.

```
            2 12
   7, 5  3̶  2̶
 - 4, 6  5   7
 ────────────────
```

Check your answer. If it is right, go to section 4.3. If it is wrong or if you don't know what to do, go to section 4.2A.

4.2A. You can subtract the units column:

```
            2 12
   7, 5  3̶  2̶
 - 4, 6  5   7
 ────────────────
               5
```

But you can't subtract the tens column.
You have to borrow from the hundreds column:

```
      4 12
        2̶ 12
   7, 5̶  3̶  2̶
 - 4, 6  5   7
 ────────────────
          7  5
```

Now you can subtract the tens column:
Finish the subtraction, check it and go to section 4.3.

4.3. Practice

Subtract and check:

```
           7 16                3 13
(a)  8, 4  8̶  6̶   (b)  9, 5  4̶  3̶   (c)  6, 7  2  3   (d) 3,541 − 1,659
   - 3, 8  8   7      - 4, 7  6   4      - 2, 7  8  4   (e) 7,325 − 4,468
   ─────────────      ─────────────      ─────────────
```

Check your answers.

5. YOU CAN'T BORROW FROM ZERO

First example: 805
              − 378

Subtract the units column . . . but 8 is more than 5. You must borrow . . . but there is a zero in the tens column. You can't borrow from zero.

So borrow from 80 instead. 80 − 1 = 79.
Put 7 in the hundreds column, 9 in the tens column.
Subtract the units column.

Subtract the tens column.
Then subtract the hundreds column.

Your answer is 427.

Check.

H	T	U
8	0	5
− 3	7	8

H	T	U
7 8	9 0	15 5
− 3	7	8
		7

H	T	U
7 8	9 0	15 5
− 3	7	8
4	2	7

H	T	U
① 4	① 2	7
+ 3	7	8
8	0	5

Second example:
          5,004
       − 1,357

Subtract the units column . . . but 7 is more than 4.
You must borrow. You can't borrow from 0. You can't borrow from 00.

So borrow from 500 instead. 500 − 1 = 499.
Put 4 in the thousands column, 9 in the hundreds column, 9 in the tens column.
Subtract the units column.

Subtract the tens column.
Subtract the hundreds column.
Subtract the thousands column.

Your answer is 3,647.

Check.

TH	H	T	U
5,	0	0	4
− 1,	3	5	7

TH	H	T	U
4 5,	9 0	9 0	14 4
− 1,	3	5	7
			7

TH	H	T	U
4 5,	9 0	9 0	14 4
− 1,	3	5	7
3,	6	4	7

TH	H	T	U
① 3,	① 6	① 4	7
+ 1,	3	5	7
5,	0	0	4

5.1. Practice

Subtract and check.

(a)
```
  2914
   304
 - 197
 ─────
```

(b)
```
  6,006
 - 2,328
 ───────
```

(c)
```
  8,002
 - 5,419
 ───────
```

(d) 705 – 526 =

(e) 401 – 239 =

6. YOU CAN SUBTRACT ANY NUMBERS

In Chapters 4 and 5, you have been learning how to subtract ANY NUMBERS. No matter how many digits they have, no matter how many times you need to borrow, all you have to do is remember the rules you have learned so far.

7. SUBTRACTION REMINDERS

Make sure that the number you subtract is not larger than the number you subtract from.

Be sure to put the number you subtract UNDER the number you subtract from.

583 – 96 =   To subtract, write in column form:
```
  583
 -  96
 ─────
```

Write units under units, tens under tens, hundreds under hundreds, thousands under thousands, and so on.

Leave some space on top of your subtraction in case you need to borrow several times.

Show all your work when you borrow.

Check by adding.

7.1. More Practice

(a)
```
  367
 -  29
 ─────
```

(b)
```
  582
 - 284
 ─────
```

(c)
```
  35
 - 26
 ────
```

(d)
```
  13
 -  8
 ────
```

(e) 252 – 61 =

(f)
```
  4,352
 - 1,078
 ───────
```

(g)
```
  6,435
 - 2,756
 ───────
```

(h)
```
  9,521
 - 2,678
 ───────
```

(i)
```
  5,030
 - 2,452
 ───────
```

(j) 93 – 15 =

Check your answers.

7.2. Word Problems

Before you start these problems, look at Chapter 4. That should remind you of the SUBTRACTION WORD CLUES. Reread HOW TO SOLVE WORD PROBLEMS in Chapter 2 if you don't remember the steps.

1. A medium size banana has 101 calories. An orange has 45 calories. How many more calories are there in a banana than in an orange?
   (1) 146   (2) 57   (3) 56   (4) 53
   (5) none of these

2. There are 37 volunteer firefighters in the village of Winsbrook Falls. 29 are men. How many women firefighters are there in this village?
   (1) 8   (2) 16   (3) 7   (4) 9
   (5) none of these

3. Atlantic Electric customers set an all-time peak demand record of 1,176,000 kilowatts on July 21. Customers used 984,300 kilowatts the next day. How much more electricity was used on the first day of the heat wave?
   (1) 391,700   (2) 191,700   (3) 290,700   (4) 190,300
   (5) none of these

4. Shopping for a home organ, the Metcalfs discovered that the least expensive cost $995 and the most expensive cost $7,650. What is the difference in price?
   (1) $6,655   (2) $6,745   (3) $7,650   (4) $6,550
   (5) none of these

5. A horseowner received $10,800 prize money when her horse won a race. After expenses, her profit was $6,849. How much were the expenses?
   (1) $3,049   (2) $4,041   (3) $3,951   (4) $3,051
   (5) none of these

Check your answers. If all your answers are right, begin Chapter 6. If one or two answers are wrong, make sure that you understand the right answers, and then go on to Chapter 6. If you made more than two mistakes, you should read Chapter 5 again. Then start Chapter 6.

8. ANSWERS

1.1. 7

1.2. (a) 9   (b) 8   (c) 9   (d) 3   (e) 9

2.1. 26; 36

2.2. (a) 38   (b) 46   (c) 39   (d) 48   (e) 29   (f) 9

      (g) 16   (h) 4

3.1. (a) 254   (b) 286   (c) 251   (d) 374   (e) 342

4.1. (a) 5,832   (b) 2,813   (c) 3,330   (d) 3,511   (e) 2,613

4.2. 2,875

4.3. (a) 4,599   (b) 4,779   (c) 3,939   (d) 1,882   (e) 2,857

5.1. (a) 107   (b) 3,678   (c) 2,583   (d) 179   (e) 162

7.1. (a) 338   (b) 298   (c) 9   (d) 5   (e) 191

      (f) 3,274   (g) 3,679   (h) 6,843   (i) 2,578   (j) 78

7.2. Word Problems

1. (3)
$$\begin{array}{r} 101 \\ -\ 45 \\ \hline 56 \end{array}$$

2. (1)
$$\begin{array}{r} 37 \\ -\ 29 \\ \hline 8 \end{array}$$

3. (2)
$$\begin{array}{r} 1,176,000 \\ -\ 984,300 \\ \hline 191,700 \end{array}$$

4. (1)
$$\begin{array}{r} \$7,650 \\ -\ 995 \\ \hline \$6,655 \end{array}$$

5. (3)
$$\begin{array}{r} \$10,800 \\ -\ 6,849 \\ \hline \$\ 3,951 \end{array}$$

SIX

Multiplication: Part I

1. WHAT IS MULTIPLICATION?

Multiplication is a faster way of adding the same number several times. If 4 lightbulbs come in a package, how many lightbulbs are there in 8 packages?

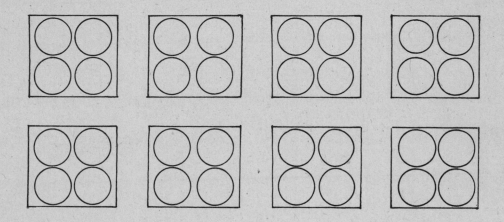

You need to know the total amount of 8 groups of 4.

Adding 4 + 4 + 4 + 4 + 4 + 4 + 4 + 4 will give you the answer: 32.

MULTIPLYING 4 × 8 is a much faster way and gives the same answer.

The two ways of writing this multiplication are:

$$4 \times 8 = 32 \qquad \text{or} \qquad \begin{array}{r} 4 \\ \times\,8 \\ \hline 32 \end{array}$$

When your answer is a two-digit number, be sure to write the units in the units column and the tens in the tens column.

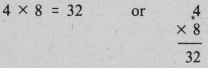

correct multiplication	incorrect multiplication
$\begin{array}{r} 4 \\ \times\,8 \\ \hline 32 \end{array}$	$\begin{array}{r} 4 \\ \times\,8 \\ \hline 32 \end{array}$

1.1. Test Yourself

Here is another example:

6 glasses on each tray

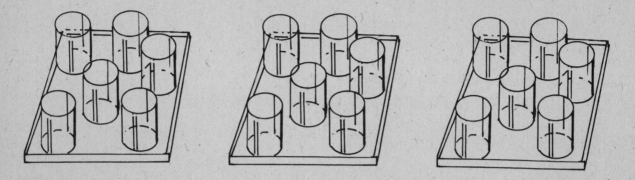

3 trays

How many glasses in all? (Write your answer in the form of a multiplication.)

Check your answer. If it is right, go to section 2. If it is wrong or if you don't know what to do, go to section 1.1A.

1.1A. You know that $6 + 6 + 6 = 18$ (or you can find it by counting the glasses one by one).

Written as a multiplication, this operation is:

$$6 \times 3 = 18 \qquad \text{or} \qquad \begin{array}{r} 6 \\ \times\, 3 \\ \hline 18 \end{array}$$

Go to section 2.

2. THE MULTIPLICATION CHART

If you learn the basic multiplication facts, you will be able to multiply more quickly. Like the addition chart you saw in Chapter 2, the multiplication chart shows the basic facts all at once.

Here is how to use the multiplication chart.

For example, let's multiply 4 × 6.

Find the number 4 in the side row.

Go straight over the row and stop under the 6.

Your answer is in the box where you stopped.

Write it down:

$$4 \times 6 = 24 \quad \text{or} \quad \begin{array}{r} 4 \\ \times 6 \\ \hline 24 \end{array}$$

×	0	1	2	3	4	5	6	7	8	9
0	0	0	0	0	0	0	0	0	0	0
1	0	1	2	3	4	5	6	7	8	9
2	0	2	4	6	8	10	12	14	16	18
3	0	3	6	9	12	15	18	21	24	27
4	0	4	8	12	16	20	24	28	32	36
5	0	5	10	15	20	25	30	35	40	45
6	0	6	12	18	24	30	36	42	48	54
7	0	7	14	21	28	35	42	49	56	63
8	0	8	16	24	32	40	48	56	64	72
9	0	9	18	27	36	45	54	63	72	81

Notice that zero multiplied by any number is ALWAYS zero.

3. HELPFUL HINTS FOR MULTIPLICATION

The TIMES SIGN is used for multiplication. It tells you to MULTIPLY the numbers you see. It looks like this: ×

When you multiply, there are two ways of writing your answer:

$$4 \times 3 = 12 \quad \text{or} \quad \begin{array}{r} 4 \\ \times 3 \\ \hline 12 \end{array}$$

The answer in multiplication is called the PRODUCT.
To check your multiplication, here is all you do:

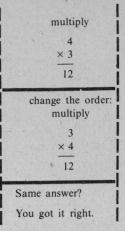

3.1. Practice

Multiply these numbers. Use the multiplication chart only if you need it. If your answer is a two-digit number, be sure to write it correctly.

(a)	**(b)**	**(c)**	**(d)**	**(e)**	**(f)** 9 × 5 =
6	3	4	5	7	**(g)** 7 × 8 =
× 4	× 5	× 0	× 1	× 3	**(h)** 6 × 9 =

Check your answers. If they are all right, go to section 4. If one or more are wrong, read sections 1.1A to 2 again and then go to section 4.

4. MULTIPLYING A TWO–DIGIT NUMBER

There are 12 doughnuts in a box of 1 dozen. How many doughnuts are there in 4 dozen?

To multiply a two-digit number, use the column form. Put units under units, and tens under tens.

Multiply the units column:

$$\begin{array}{r} 2 \\ \times\,4 \\ \hline 8 \end{array}$$

Then multiply the tens column:

$$\begin{array}{r} 1 \\ \times\,4 \\ \hline 4 \end{array}$$

Your answer is 48.

There are 48 doughnuts in 4 dozen.

4.1. Practice

Be sure to multiply both the units column and the tens column. If a multiplication is not in the column form, put it in the column form first.

    (a)   14   **(b)**   23   **(c)**   54   **(d)** $23 \times 2 =$
        $\times\ 2$ $\times\ 3$ $\times\ 1$ **(e)** $33 \times 3 =$

Check your answers. If they are all right, go to section 5. If one or more are wrong, read section 4 again, and then go to section 5.

5. CARRYING OVER

Sometimes you need to carry over while you are doing a multiplication.
Here is an example: 29×2

Put your multiplication in the column form.
Multiply units column: $9 \times 2 = 18$.
CARRY the 1 from the 18 OVER to the top of the tens column.

Multiply the tens column: $2 \times 2 = 4$.
ADD the number you carried over: $4 + 1 = 5$.

Your answer is 58.

5.1. Test Yourself

Do this multiplication: 24
                          $\times\ 3$

Check your answer. If it is right, go to section 5.2. If it is wrong or if you don't know what to do, go to section 5.1A.

5.1A. Follow the same steps as in the example given before.

Multiply units: $4 \times 3 = 12$
Carry the 1 from the 12 over to the top of the tens column.
Multiply tens: $2 \times 3 = 6$.
Add the number you carried over: $6 + 1 = 7$.
Your answer is 72.

Important: The number you carried over is ALWAYS ADDED, never multiplied.

Sometimes you have to carry over a number greater than 1.

    Example: 28 × 3

Put your multiplication in the column form.

Multiply units column: 8 × 3 = 24.

T	U
②2	8
×	3
	4

Carry the 2 from the 24 over to the top of the tens column.

Multiply the tens column: 2 × 3 = 6.

Add the number you carried over: 6 + 2 = 8.

Your answer is 84.

T	U
②2	8
×	3
8	4

5.3. Practice

(a) ① 12 × 5   (b) 39 × 2   (c) 18 × 4   (d) 27 × 3 =   (e) 25 × 3 =

Check your answers. If they are all right, go to section 6. If one or more are wrong, read sections 4 and 5 again, read section 5.3 and the go to section 6.

5.4. Sometimes you need to carry over from the tens column to the hundreds column. Here is an example.

    63 × 2

Put your multiplication in the column form.

Multiply units: 3 × 2 = 6.

H	T	U
	6	3
×		2
		6

Multiply the tens column: 6 × 2 = 12

The 1 is carried over to the hundreds column.

H	T	U
	6	3
×		2
①	2	6

When you start a multiplication, always leave some space on the left side. You may need a new column to carry over.

6. CARRYING OVER MORE THAN ONCE

Carrying over can happen in more than one column.

    46 × 4

Put your multiplication in the column form.
Multiply units: $6 \times 4 = 24$
Carry the 2 from the 24 over to the top of the tens column.

	H	T	U
		②4	6
×			4
			4

Multiply tens: $4 \times 4 = 16$.
Add the number you carried over: $16 + 2 = 18$.
Carry the 1 to the hundreds column.

	H	T	U
		②4	6
×			4
	①	8	4

Your answer is 184.

6.1. Test Yourself

Do this multiplication: 39
                         × 5

Check your answer. If it is right, go to section 6.3. If it is wrong or if you don't know what to do, go to section 6.1A.
6.1A. Follow the same steps as in the example given before.
Multiply units: $9 \times 5 = 45$. Carry the 4 over to the top of the tens column.

	H	T	U
		④3	9
×			5
	①	9	5

Multiply tens: $5 \times 3 = 15$.
Add the number carried over: $15 + 4 = 19$.
Carry the 1 over to hundreds column.

Your answer is 195.

6.3. Practice

For these multiplications, you may have to carry over in one column only, or in two columns. Leave some space on the left side in case you need a new column.

   (a)   56   **(b)**   37   **(c)**   92   **(d)** $36 \times 4 =$
       × 2   × 3   × 2   **(e)** $51 \times 3 =$

Check your answers. If they are all right, go to section 7. If one or more are wrong, read sections 5.3 and 6 again, and then go to section 7.

7. MULTIPLYING A THREE-DIGIT NUMBER

Example: 424×2
Put the multiplication in the column form.

Multiply units: $4 \times 2 = 8$.

Multiply tens: $2 \times 2 = 4$.

Multiply hundreds: $4 \times 2 = 8$.

Your answer is 848.

	H	T	U
	4	2	4
×			2
			8

	H	T	U
	4	2	4
×			2
		4	8

	H	T	U
	4	2	4
×			2
	8	4	8

7.1. Practice

(a)   321
       × 3

(b)   402
       × 2

(c)   222
       × 4

(d) $999 \times 1 =$

(e) $673 \times 0 =$

Check your answers.

8. CARRYING OVER

When you multiply a three-digit number, you may have to carry over. It may be in one column only, or in more than one column. You may even need a new column on the left side of your multiplication: the thousands column.

    228×7
Put the multiplication in the column form.
Multiply units: $8 \times 7 = 56$.
Carry the 5 from 56 over to the top of the tens column.

Multiply the tens column: $2 \times 7 = 14$
Add the number you carried over: $14 + 5 = 19$.
Carry the 1 over to the top of the hundreds column.

Multiply the hundreds column: $2 \times 7 = 14$.
Add the number you carried over: $14 + 1 = 15$.
Carry the 1 from the 15 to the thousands column. Your answer is 1,596.

TH	H	T	U
	2	⑤2	8
×			7
			6

TH	H	T	U
	①2	⑤2	8
×			7
		9	6

TH	H	T	U
	①2	⑤2	8
×			7
①	5	9	6

8.1. Practice

(a) 273   (b) 157   (c) 258   (d) 376 × 2 =
    × 3   × 4   × 3   (e) 564 × 3 =

Check your answers.

8.2. Practice with Longer Numbers

(a) 6,425   (b) 8,452   (c) 2,234   (d) 4,923 × 4 =
    × 3   × 4   × 2   (e) 7,567 × 6 =

Check your answers. If they are all right, go to section 9. If not, read sections 7 and 8 again. Then go to section 9.

9. MULTIPLYING MORE THAN TWO NUMBERS

Try to multiply these numbers all at once:

2 × 5 × 3

Can't do it? No. It's impossible to multiply more than two numbers at a time! Here is what you have to do:

Multiply the two first numbers:

Then multiply the product by the last number:

Your answer is 30.

9.1. Practice

(a) 1 × 2 × 3 × 4 =   (b) 5 × 4 × 2 =   (c) 3 × 3 × 0 × 9 =

(d) 3 × 2 × 6 =   (e) 4 × 4 × 2 × 2 =   Check your answers.

9.2. More Practice

(a)   12   (b)   46   (c)   252   (d)   704   (e)   35
     × 3   × 5   ×   3   ×   6   × 5
    ─────   ─────   ──────   ──────   ─────

Check your answers.

10. WORD PROBLEMS—MULTIPLICATION

MULTIPLICATION is a faster way of adding the same number several times. Multiplication word problems are easier to solve if you look for multiplication word clues. Some multiplication word clues are:

(a) TIMES—15,675 people went to the fireworks display in 1975. Six times that amount went in 1976. How many people saw the fireworks in 1976?

(b) AT—What is the price of eight shirts selling at 9 dollars each?

(c) TOTAL—48 cars park in one row of a parking lot at a stadium. What is the total number of cars parked in nine rows?

(d) IN ALL—In an apartment building there are 14 steps between floors. If there are six floors, how many steps in all are there to walk?

TIMES—   15,675
          ×   6
          ─────────
            94,050   94,050 people saw the fireworks in 1976.

AT—   8
        × $9
        ─────
        $72   The cost is $72.

TOTAL—   48
         × 9
         ──────
            432   432 cars are in 9 rows.

IN ALL—   14
          × 6
          ──────
             84   There are 84 steps in all.

10.1. Word Problems

1. A cinder block weighs 22 pounds. How many pounds do 5 cinder blocks weigh?
   (1) 100   (2) 110   (3) 101   (4) 120
   (5) none of these

    1.   1   2   3   4   5

2. A pilot flies 2,658 miles for 4 days. How many miles in all does the pilot fly?
   (1) 9,632   (2) 10,432   (3) 10,632   (4) 10,300
   (5) none of these

    2.   1   2   3   4   5

3. The farmer planted 48 rows of apple trees with 9 trees in each row. How many apple trees will there be in the apple orchard?
   (1) 422   (2) 367   (3) 424   (4) 432
   (5) none of these

    3.   1   2   3   4   5

4. A father weighs 6 times as much as his child. The child weighs 35 pounds. How much does the father weigh?
   (1) 180   (2) 230   (3) 210   (4) 181
   (5) none of these

    4.   1   2   3   4   5

5. There were 358 crates of lamps on a loading dock. Each crate contained 8 lamps. What was the total number of lamps?
   (1) 2,864   (2) 2,888   (3) 2,486   (4) 2,600
   (5) none of these

    5.   1   2   3   4   5

6. Ricardo needs 2 wall braces for each shelf he puts up. He puts up 8 shelves. How many wall braces does he need all together?
   (1) 14   (2) 16   (3) 18   (4) 10
   (5) none of these

    6.   1   2   3   4   5

7. A rock band gave two concerts a night for 3 nights. At each concert there were 1,000 people. How many people saw the concert?
   (1) 2,000   (2) 5,000   (3) 2,300   (4) 6,000
   (5) none of these

    7.   1   2   3   4   5

8. 8,195 people each bet 2 dollars on the first horse race at Belmont. What was the total amount of money on the first race?
   (1) $16,280   (2) $17,930   (3) $16,390   (4) $16,000
   (5) none of these

    8.   1   2   3   4   5

11. ANSWERS

1.2. 18

3.1. (a) 24   (b) 15   (c) 0   (d) 5   (e) 21   (f) 45

       (g) 56   (h) 54

4.1. (a) 28   (b) 69   (c) 54   (d) 46   (e) 99

5.1. 72

5.2. (a) 60   (b) 78   (c) 72   (d) 81   (e) 75

6.1. 195

6.2. (a) 112   (b) 111   (c) 184   (d) 144   (e) 153

7.1. (a) 963   (b) 804   (c) 888   (d) 999   (e) 0

8.1. (a) 819   (b) 628   (c) 774   (d) 752   (e) 1,692

8.2. (a) 19,275   (b) 33,808   (c) 4,468   (d) 19,692   (e) 45,402

9.1. (a) 24   (b) 40   (c) 0   (d) 36   (e) 64

9.2. (a) 36   (b) 230   (c) 756   (d) 4,224   (e) 175

10.1. Word Problems

1. (2)
$$\begin{array}{r} 22 \\ \times\ 5 \\ \hline 110 \end{array}$$

2. (3)
$$\begin{array}{r} 2,658 \\ \times\ \ \ \ 4 \\ \hline 10,632 \end{array}$$

3. (4)
$$\begin{array}{r} 48 \\ \times\ 9 \\ \hline 432 \end{array}$$

4. (3)
$$\begin{array}{r} 35 \\ \times\ 6 \\ \hline 210 \end{array}$$

5. (1)
$$\begin{array}{r} 358 \\ \times\ \ 8 \\ \hline 2,864 \end{array}$$

6. (2)
$$\begin{array}{r} 2 \\ \times\ 8 \\ \hline 16 \end{array}$$

7. (4) $2 \times 3 \times 1,000 = 6,000$

8. (3)
$$\begin{array}{r} 8,195 \\ \times\ \ \$2 \\ \hline \$16,390 \end{array}$$

SEVEN

Multiplication: Part II

1. MULTIPLYING A TWO-DIGIT NUMBER

A basketball superstar shot 42 points for 23 games in a row. What was the total number of points he scored?

Of course, we are not going to add 42 twenty-three times. It's much easier to multiply: 42×23. Here is how to do it.

Put multiplication in column form. Put another column on the left side: you may need it to carry over.
First multiply $42 \times 3 \ldots 126$.

	H	T	U
		4	2
×		2	3
	1	2	6

Then multiply 42×2.
Begin your answer in the tens column because the 2 you multiply by is in the tens column.
Put a zero in the units column to keep the row in place.
Answer: 840.

		4	2
×		2	3
	1	2	6
	8	4	0

Add the two numbers:

```
  126
+ 840
-----
  966
```

		4	2
×		2	3
	1	2	6
	8	4	0
	9	6	6

Your answer is 966.
The basketball player scored 966 points altogether.

1.1. Test Yourself

Do this multiplication:

$$\begin{array}{r} 16 \\ \times\ 12 \\ \hline \end{array}$$

Check your answer. If it is right, go to section 1.2. If it is wrong or you don't know what to do, go to section 1.1A.

1.1A. First, multiply 16×2 (you need to carry over).
Answer: 32.

Then multiply 16×1.
Begin your answer under the 1 in the tens column.
Put a zero in the units column. Answer: 160.
Add the two numbers. Your answer is 192.

H	T	U
	①	
	1	6
×	1	2
	3	2
1	6	0
1	9	2

1.2. You may have to carry over, not only when you multiply, but also when you add your two numbers to find your final answer.

$$37 \times 45$$

Put multiplication in column form.
Multiply 37×5. Answer: 185 (you need to carry over).

TH	H	T	U
		③	
		3	7
×		4	5
	1	8	5

Multiply 37×4.
Begin your answer in the tens column, under the 4 (you need to carry over).
Put a 0 in the units column. Answer: 1480.

TH	H	T	U
		②	
		③	
		3	7
×		4	5
	1	8	5
1	4	8	0

Add the two numbers you got (you need to carry over).
Your final answer is 1,665.

TH	H	T	U
		②	
		③	
		3	7
×		4	5
	①1	8	5
1	4	8	0
1,	6	6	5

1.3. Practice

(a)
$$\begin{array}{r} 27 \\ \times\ 50 \\ \hline 00 \\ 1350 \\ \hline \end{array}$$

(b)
$$\begin{array}{r} 44 \\ \times\ 23 \\ \hline \end{array}$$

(c)
$$\begin{array}{r} 92 \\ \times\ 27 \\ \hline \end{array}$$

(d) $76 \times 48 =$
(e) $36 \times 53 =$

Check your answers.

2. MULTIPLYING A THREE-DIGIT NUMBER BY A TWO-DIGIT NUMBER

The steps are exactly the same as in section 1.

327×54

Put in column form, leaving some space on the left side.
Multiply 327×4. Answer: 1308.

T TH	TH	H	T	U
		①	②	
		3	2	7
×			5	4
	1	3	0	8

Multiply 327×5.
Start writing in the tens column.
Put a 0 in the units column.
Answer: 16350.

T TH	TH	H	T	U
		①	③	
		①	②	
		3	2	7
×			5	4
	1	3	0	8
1	6	3	5	0

Add the two numbers you got.
Final answer: 17,658

T TH	TH	H	T	U
		①	③	
		①	②	
		3	2	7
×			5	4
	1	3	0	8
1	6	3	5	0
1	7,	6	5	8

You could also multiply a four-digit number, or a five-digit number, or any number by a two-digit number.

2.1. Practice

(a) 649
× 38

(b) 4,252
× 17

(c) 26,407
× 39

(d) $734 \times 68 =$
(e) $9,523 \times 54 =$

Check your answers.

3. MULTIPLYING A THREE-DIGIT NUMBER BY A THREE-DIGIT NUMBER

When you multiply by a three-digit number, you will have 3 rows of adding.

357×249

Put in column form (leaving a lot of space on the left side).
Multiply 357×9. Answer: 3213

T TH	TH	H	T	U
		⑤	⑥	
		3	5	7
×		2	4	9
	3	2	1	3

Multiply 357 × 4.

Begin your answer in the tens column.
Put a 0 in the units column.
Answer: 14280.

Multiply 357 × 2.
Begin your answer in the hundreds column.
Put a 0 in the units column, and a 0 in the tens column.
Answer: 71400.

Add the three numbers you got.
Final answer: 88,893.

```
         ②②
         ⑤⑥
          3  5  7
    ×     2  4  9
    ─────────────
       3  2  1  3
    1  4  2  8  0
```

```
         ①①
         ②②
         ⑤⑥
          3  5  7
    ×     2  4  9
    ─────────────
       3  2  1  3
    1  4  2  8  0
    7  1  4  0  0
```

```
         ①①
         ②②
         ⑤⑥
          3  5  7
    ×     2  4  9
    ─────────────
       3  2  1  3
    1  4  2  8  0
    7  1  4  0  0
    ─────────────
    8  8, 8  9  3
```

3.1. You could also multiply a four-digit number or a five-digit number by a three-digit number.
You would always have 3 rows of adding, with a zero in the second row and two zeros in the third row.

3.2. Practice

    (a)   725   **(b)**   4,354   **(c)**   9,412   **(d)** 543 × 55 =
         × 124   × 372   × 46   **(e)** 2,945 × 143 =

Check your answers.

3.3. You could also multiply by a four-digit number. You would have 4 rows of adding, with a zero in the second, two zeros in the third row, and three zeros in the fourth row.
Here is an example of the multiplication of a five-digit number by a four-digit number.
We have not written the numbers carried over.

$$
\begin{array}{r}
35{,}742 \\
\times\; 6{,}431 \\
\hline
35\ 742 \\
1\ 072\ 260 \\
14\ 296\ \mathbf{800} \\
214\ 452\ \mathbf{000} \\
\hline
229{,}856{,}802
\end{array}
$$

4. MULTIPLYING BY 10, 100 or 1,000—A SHORT WAY

To multiply any number by 10, add a zero on the right side.

$$\begin{array}{r} 497 \\ \times\ 10 \\ \hline 000 \\ 4970 \\ \hline 4,970 \end{array}$$

Long way

$$\begin{array}{r} 497 \\ \times\ 10 \\ \hline 4,970 \end{array}$$

Short way

To multiply any number by 100, add 00 on the right side.

$$\begin{array}{r} 8,327 \\ \times\ 100 \\ \hline 0000 \\ 00000 \\ 832700 \\ \hline 832,700 \end{array}$$

Long way

$$\begin{array}{r} 8,327 \\ \times\ 100 \\ \hline 832,700 \end{array}$$

Short way

To multiply any number by 1,000, add 000 on the right side.

$$\begin{array}{r} 73 \\ \times\ 1,000 \\ \hline 00 \\ 000 \\ 0000 \\ 73000 \\ \hline 73,000 \end{array}$$

Long way

$$\begin{array}{r} 73 \\ \times\ 1,000 \\ \hline 73,000 \end{array}$$

Short way

The short way always works. It saves time and space.

4.1. Practice

Multiply the short way

(a) $\begin{array}{r} 947 \\ \times\ 10 \\ \hline \end{array}$ (b) $\begin{array}{r} 32 \\ \times\ 100 \\ \hline \end{array}$ (c) $\begin{array}{r} 6,744 \\ \times\ 1,000 \\ \hline \end{array}$ (d) $27 \times 100 =$
(e) $4,935 \times 10 =$

Check your answers.

5. YOU CAN MULTIPLY ANY NUMBERS

In Chapters 6 and 7, you have been learning how to multiply ANY NUMBERS. No matter how many digits they have, no matter how many of them you need to multiply, no matter how many times you need to carry over, all you have to do is remember the rules you have learned so far.

6. MULTIPLICATION REMINDERS

Be sure to put the numbers you are multiplying in the correct columns before you multiply.

428 × 34   To multiply, write in
                column form
                  428
                × 34
                ─────

Put units under units, tens under tens, hundreds under hundred, thousands under thousands.

Leave some space on the left side of your multiplications in case you need new columns to carry over.

Be sure to put the zeros in place when you multiply by 2 or more digits.

To get the correct final answer, multiply each digit carefully and add your rows carefully.

Remember the short way to multiply by 10, 100, or 1,000.

6.1. More Practice

(a)   33   (b)   46   (c) 7,527   (d)   289
     × 32   × 13   ×   48   × 624
     ─────   ─────   ──────   ──────

(e) 3,876   (f)   937   (g)   37   (h)   252
   × 2,043   × 74   × 68   × 76
   ───────   ─────   ─────   ─────

(i) 2,893   (j)   597
   ×   36   × 214
   ───────   ─────

Check your answers.

6.2. Word Problems

Before you start, look at Chapter 6. That will remind you of the MULTIPLICATION WORD CLUES. Read HOW TO SOLVE WORD PROBLEMS in Chapter 2 to remind you of the steps to follow.

1. There are 12 elevators in a large office building. Each elevator holds a maximum of 23 people. How many people can ride in all the elevators at once?
   (1) 69   (2) 2,706   (3) 276   (4) 267
   (5) none of these

   1. 1 2 3 4 5

2. You have two loan payments a month. Each loan payment is 56 dollars. You have 21 more payments. How much money do you owe?
   (1) $2,352   (2) $1,176   (3) $2,450   (4) $2,300
   (5) none of these

   2. 1 2 3 4 5

3. A cruise ship carries 948 passengers. How many passengers are on 20 ships?
   (1) 18,000   (2) 1,896   (3) 18,960   (4) 19,960
   (5) none of these

   3. 1 2 3 4 5

4. The unemployment office interviews 148 people every day. How many people do they interview in 21 days?
   (1) 3,100   (2) 3,108   (3) 2,108   (4) 2,008
   (5) none of these

   4. 1 2 3 4 5

5. A group of 368 people ate 3 meals a day for 14 days at a hotel. How many meals were prepared in 14 days?
   (1) 1,156   (2) 11,456   (3) 15,115   (4) 15,456
   (5) none of these

   5. 1 2 3 4 5

6. A group of 46 volunteers are out to collect money to send needy children to camp. If each volunteer visits 35 homes a day for 3 days, how many homes in all will be visited?
   (1) 1,156   (2) 4,300   (3) 14,480   (4) 4,830
   (5) none of these

   6. 1 2 3 4 5

7. A truck carries 135 cases of soda. There are 24 cans to a case. How many cans of soda are there?
   (1) 3,240   (2) 3,940   (3) 2,340   (4) 3,205
   (5) none of these

   7. 1 2 3 4 5

8. The Benson's have 3 more years to pay off the mortgage on their home. They pay $150 a month. How much more money do they owe the bank? (There are 12 months in a year.)
   (1) $4,500   (2) $1,800   (3) $5,400   (4) $4,500
   (5) none of these

   8. 1 2 3 4 5

Check your answers. If all your answers are correct, start Chapter 8. If one, two, or three answers are wrong, make sure that you understand the right answers. Then start Chapter 8. If more than three answers are wrong, you should read Chapter 7 again before starting Chapter 8.

7. ANSWERS

1.1.
$$\begin{array}{r} 16 \\ \times\ 12 \\ \hline 32 \\ 160 \\ \hline 192 \end{array}$$

1.3. (a) 1,350   (b) 1,012   (c) 2,484   (d) 3,648   (e) 1,908

2.1. (a) 24,662   (b) 72,284   (c) 1,029,873   (d) 49,912   (e) 514,242

3.2. (a) 89,900   (b) 1,619,688   (c) 432,952   (d) 29,865   (e) 421,135

4.1. (a) 9,470   (b) 3,200   (c) 6,744,000   (d) 2,700   (e) 49,350

6.1. (a) 1,056   (b) 598   (c) 361,296   (d) 180,336   (e) 7,918,668

     (f) 69,338   (g) 2,516   (h) 19,152   (i) 104,148   (j) 127,758

6.2. Word Problems

1. (3)
$$\begin{array}{r} 23 \\ \times\ 12 \\ \hline 46 \\ 230 \\ \hline 276 \end{array}$$

2. (1) $2 \times 56 \times 21 =$

$$\begin{array}{r} \$56 \\ \times\ 2 \\ \hline \$112 \end{array} \qquad \begin{array}{r} \$112 \\ \times\ 21 \\ \hline 112 \\ 2240 \\ \hline \$2,352 \end{array}$$

3. (3)
$$\begin{array}{r} 948 \\ \times\ 20 \\ \hline 18,960 \end{array}$$

4. (2)
$$\begin{array}{r} 148 \\ \times\ 21 \\ \hline 148 \\ 2960 \\ \hline 3,108 \end{array}$$

5. (4) $14 \times 3 \times 368 =$

$$\begin{array}{r} 14 \\ \times\ 3 \\ \hline 42 \end{array} \qquad \begin{array}{r} 368 \\ \times\ 42 \\ \hline 736 \\ 14\ 720 \\ \hline 15,456 \end{array}$$

6. (4) $46 \times 35 \times 3 =$

$$\begin{array}{r} 35 \\ \times\ 3 \\ \hline 105 \end{array} \qquad \begin{array}{r} 105 \\ \times\ 46 \\ \hline 630 \\ 4200 \\ \hline 4,830 \end{array}$$

7. (1)
$$\begin{array}{r} 135 \\ \times\ 24 \\ \hline 540 \\ 2700 \\ \hline 3,240 \end{array}$$

8. (3) $12 \times 3 \times \$150 =$

$$\begin{array}{r} 12 \\ \times\ 3 \\ \hline 36 \end{array} \qquad \begin{array}{r} \$150 \\ \times\ 36 \\ \hline 900 \\ 4500 \\ \hline \$5400 \end{array}$$

EIGHT

Division: Part I

1. WHAT IS DIVISION?

Finding out how many times one number is contained in another number is called DIVISION.

You have 12 eggs. How many people will have 2 eggs each?

You can see that there are 6 groups of 2 in 12. So 6 people will have 2 eggs each. There are two ways of writing this division.

You can write it either this way:

$$12 \div 2 = 6 \qquad \text{(you say: 12 divided by 2 is 6)}$$

or this way:

$$2 \overline{)\,12\,}^{\,6} \qquad \text{(you say: 2 goes into 12, 6 times)}$$

1.1. Test Yourself

You have a bunch of 15 flowers. You want to make bunches of 3 flowers to put in small vases. How many vases will you need?

Check your answer. If it is right, go to section 1.2. If it is wrong or if you don't know what to do, go to section 1.1A.

1.1A. To find out, make groups of 3 flowers until you have no flowers left:

There are 5 groups of 3 flowers, so the answer is 5.

79

1.2. The two ways of writing your division are:

$$15 \div 3 = 5 \qquad \text{or} \qquad 3 \overline{)\ 15}^{\ 5}$$

2. DIVISION IS RELATED TO MULTIPLICATION

John, Carlos, Maria, and Jane are playing cards. The 32 cards are distributed to the players so that each player gets the same amount of cards. Each player gets:

$$4 \overline{)\ 32}^{\ 8} \qquad \text{8 cards}$$

At the end of the game, the players put their cards back on the table. The pack of cards has:

$$8 \times 4 = 32 \quad \text{cards in it.}$$

3. HELPFUL HINTS FOR DIVISION

Two signs tell you to divide.
One division sign looks like this:

$$15 \div 3$$

The number on the LEFT of the division sign is divided by the number on the RIGHT.
"15 divided by 3 is how much?"
The other division sign looks like this:

$$3 \overline{)\ 15}$$

The number outside the sign goes into the number inside the sign: "3 goes into 15 how many times?"
Use this sign when you divide.
Some people use other signs; for example, this one:

$$15 \mid \underline{3}$$

If you know this sign already, you can use it to divide. You don't need to learn the other one.
The answer in division is called the QUOTIENT.
To check your division, here is all you do:

$$3 \overline{)\ 15}^{\ 5} \qquad \text{divide}$$

$$\begin{array}{r} 5 \\ \times\ 3 \\ \hline 15 \end{array} \qquad \text{multiply}$$

Find the number inside the sign?

You got it right.

3.1. Practice

Divide and check:

(a) 5 $\overline{)\,25}$ **(b)** 6 $\overline{)\,18}$ **(c)** 3 $\overline{)\,15}$ **(d)** 27 ÷ 9 =   **(e)** 49 ÷ 7 =

Check your answers. If they are all right, go to section 4. If one is wrong or if you don't know what to do, go to section 3.1A.

3.1A. Don't be confused by the zeros.

4 $\overline{)\,20}$

4 goes into 20, 5 times: the answer is 5.

Check: 5 × 4 = 20

0 ÷ 4 =

If you have 0 dollars in your pocket and you want to give them to 4 friends so that each friend gets the same amount of money, you say, "Sorry." Each of them receives 0 dollars, that is, nothing at all.

So 0 ÷ 4 = 0

For the other divisions, you may find it helpful to make drawings. For example, here is a drawing you can make for 28 ÷ 7

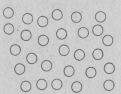   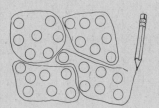

28 objects   make as many groups   there are 4 groups
                  of 7 as you can

So 28 ÷ 7 = 4

Try to remember these DIVISION FACTS so that you can divide more quickly. When you check by multiplying, you can use the MULTIPLICATION CHART in Chapter 6.

4. YOU CAN'T DIVIDE BY ZERO

It is possible to divide zero by another number, but it is IMPOSSIBLE TO DIVIDE A NUMBER BY ZERO.

Look: $\dfrac{?}{0\,)\,5}$ or $\dfrac{\overset{?}{\times 0}}{5}$ There is no answer for this.

4.1. Practice

Some of these divisions are possible, but some are impossible. Do the possible ones.

(a) $6 \overline{)\ 0}$ (b) $3 \overline{)\ 0}$ (c) $0 \overline{)\ 4}$ (d) $0 \overline{)\ 6}$ (g) $0 \div 7 =$
                                                                                    (h) $0 \div 9 =$
                                                                                    (i) $9 \div 0 =$
(e) $0 \overline{)\ 3}$ (f) $5 \overline{)\ 0}$ (j) $10 \div 0 =$

Check your answers. If they are not all correct, read section 3.1A and 4 again. Then go to section 5.

5. DIVIDING A TWO-DIGIT NUMBER

You buy 3 cans of tomatoes for 96 cents. How much do you pay for each can?
     $96 \div 3$

To divide a two-digit number, use the $\overline{)\ }$ division sign.

$$\begin{array}{c} \text{T U} \\ 3\,\overline{)\,9\ 6} \end{array}$$

First, divide the tens column. 3 goes into 9 three times. Write the answer over the 9.

$$\begin{array}{c} 3 \\ 3\,\overline{)\,9\ 6} \end{array}$$

Divide the units column. 3 goes into 6 two times. Write the answer over the 6.

$$\begin{array}{c} 3\ 2 \\ 3\,\overline{)\,9\ 6} \end{array}$$

Your answer is 32.
Each can costs $.32.

Check

$$\begin{array}{r} 32 \\ \times\ 3 \\ \hline 96 \end{array}$$

5.1. Each time you do a division, you always start from the column on the left. Imagine that you won the lottery and you want to share the prize money with a friend. The prize is several million dollars. Are you going to share the cents first? Always start divisions from the largest column.

5.2. Test Yourself

Do this division: $3 \overline{)\ 69}$
Check it by multiplying.
Check your answer. If it is right, go to section 5.3. If it is wrong or you don't know what to do, go to section 5.2A.

5.2A. You start your division from the left side (the tens column). 3 has to go into 6. It goes 2 times. Write the answer over the 6:

$$\frac{2}{3\,)\,\overline{69}}$$

Now you can finish this division. Check it and go to section 5.3.

5.3. Practice

Divide and check. Use the correct division sign.

(a) $2\,)\,\overline{48}$ **(b)** $4\,)\,\overline{84}$ **(c)** $6\,)\,\overline{66}$ **(d)** $55 \div 5 =$
                                               (e) $63 \div 3 =$

Check your answers.

6. DIVIDING A THREE-DIGIT NUMBER

Example: $684 \div 2$

Use the other division sign.
Divide the hundreds column: 2 goes into 6 three times.
Write 3 in the hundreds column.

```
  H T U
    3
2 ) 6 8 4
```

Divide the tens column.
2 goes into 8 four times. Write 4 in the tens column.

```
    3 4
2 ) 6 8 4
```

Divide the units column. 2 goes into 4 two times. Write 2 in the units column.

```
    3 4 2
2 ) 6 8 4
```

Your answer: 342.

Check.

```
    3 4 2
  ×     2
    6 8 4
```

6.1. If there is a zero in the number you divide, remember that any number goes into zero, zero times. Put the zero in its place.
Example: $804 \div 4$

Use the other division sign.
Divide the hundreds column: 4 goes into 8 two times.
Write 2 in the hundreds column.

```
    2
4 ) 8 0 4
```

Divide the tens column: 4 goes into 0 zero times.
Write 0 in the tens column.

Divide the units column. 4 goes into 4 one time.
Write 1 in the units column.

Your answer: 201.

Check.

$$\begin{array}{r} 2\ 0\ \ \\ 4\,\overline{)\,8\ 0\ 4} \end{array}$$

$$\begin{array}{r} 2\ 0\ 1 \\ 4\,\overline{)\,8\ 0\ 4} \end{array}$$

$$\begin{array}{r} 2\ 0\ 1 \\ \times\quad 4 \\ \hline 8\ 0\ 4 \end{array}$$

6.2. Practice

Divide and check. Use the correct division sign

(a) $3\,\overline{)\,639}$ (b) $4\,\overline{)\,848}$ (c) $2\,\overline{)\,642}$ (d) $5\,\overline{)\,505}$ (e) $660 \div 6 =$
(f) $846 \div 2 =$
(g) $484 \div 4 =$
(h) $707 \div 7 =$

Check your answers.

7. DIVIDING A FOUR–DIGIT NUMBER

During rush hour, 8,462 cars passed through two toll gates.
How many cars passed through each gate?
$$8,462 \div 2$$
Use the other division sign.
Start from the left, from the thousands column.
Divide the thousands column. 2 goes into 8, four times. Put 4 in
the thousands column.

Divide the hundreds column.

Divide the tens column.

Divide the units column.

Your answer: 4,231.
4,231 cars passed through each gate.

Check.

TH	H	T	U

$$\begin{array}{r} 4\quad\ \ \\ 2\,\overline{)\,8,\ 4\ 6\ 2} \end{array}$$

$$\begin{array}{r} 4\ 2\quad\ \ \\ 2\,\overline{)\,8,\ 4\ 6\ 2} \end{array}$$

$$\begin{array}{r} 4\ 2\ 3\ \\ 2\,\overline{)\,8,\ 4\ 6\ 2} \end{array}$$

$$\begin{array}{r} 4\ 2\ 3\ 1 \\ 2\,\overline{)\,8,\ 4\ 6\ 2} \end{array}$$

$$\begin{array}{r} 4,\ 2\ 3\ 1 \\ \times\qquad 2 \\ \hline 8,\ 4\ 6\ 2 \end{array}$$

7.1. Practice

Divide and check. Use the right division sign.

(a) 3) 9,366   **(b)** 4) 8,448   **(c)** 3) 6,936   **(d)** 4,242 ÷ 2 =
                                                                              (e) 5,550 ÷ 5 =

Check your answers.

7.2. Joining Two Columns

Sometimes, you cannot divide the first column when you start your division. If this happens, JOIN the first two columns.

Example: 368 ÷ 4

Use the other division sign.
Start from the left, from the hundreds column.
Divide the hundreds column . . . but 4 doesn't go into 3.

JOIN the 3 and the 6. Think of 36 in the tens column. Divide the tens column. 4 goes into 36 nine times. Write 9 in the tens column.

Divide the units column. 4 goes into 8 two times. Write 2 in the units column. Your answer: 92.

Check.

7.3. Test Yourself

Do this division: 255 ÷ 5
Check your answer. If it is right, go to section 8. If it is wrong or you don't know what to do, go to section 7.3A.
7.3A. Use the other division sign to write your division:

5) 255

You can't divide the hundreds column because 5 doesn't go into 2.

JOIN the hundreds column with the tens column. Think of 25 in the tens column.
Divide the tens column. 5 goes into 25 five times:

$$\begin{array}{r} 5 \\ 5\,\overline{)\,255} \end{array}$$

Now you can finish the division. Check it and go to section 8.

8. ZERO IN THE QUOTIENT

Sometimes you can divide the first column, but the number in the next column is two small. If this happens, you join two columns.

Example: 627 ÷ 3

Use the other division sign.
Divide the hundreds column. 3 goes into 6 two times.
Write 2 in the hundreds column.

Divide the tens column . . . but 3 doesn't go into 2.
Put zero in the tens column (3 goes
into 2 zero times).

JOIN the 2 and the 7. Think of 27 in the units column.
Divide the units column. 3 goes into 27 nine times.
Write 9 in the units column. Your answer: 209

Check.

$$\begin{array}{c} \text{H T U} \\ 2 \\ 3\,\overline{)\,6\ 2\ 7} \end{array}$$

$$\begin{array}{c} 2\ 0 \\ 3\,\overline{)\,6\ 2\ 7} \end{array}$$

$$\begin{array}{c} 2\ 0\ 9 \\ 3\,\overline{)\,6\ 2\ 7} \end{array}$$

$$\begin{array}{r} 2\ \textcircled{0}\ 9 \\ \times\quad 3 \\ \hline 6\ 2\ 7 \end{array}$$

8.1. Practice

Divide and check.

(a) $3\,\overline{)\,276}$ (b) $2\,\overline{)\,214}$ (c) $4\,\overline{)\,432}$ (d) $6\,\overline{)\,426}$ (e) 520 ÷ 5 =
(f) 546 ÷ 6 =
(g) 615 ÷ 3 =
(h) 369 ÷ 9 =

Check your answers.

9. WHAT TO DO WITH LEFTOVERS

No, this is not a cookbook! You get leftovers in division when you cannot divide an amount equally.

Can three people share 10 subway tokens equally?

No. There is one token left over. In division, the 1 left over is called the REMAINDER.

3 goes into 10 three times, but 3 × 3 is only 9.

There is a remainder: 1

$$3 \overline{)1\ 0} \quad \begin{array}{r} 3 \\ \end{array}$$

$$3 \overline{)1\ 0} \quad \begin{array}{r} 3 \text{ r1} \\ \end{array}$$

The remainder is always LESS than the number you divided by. To check your division, here is what to do:

Multiply   3
          × 3
          ———
             9

Add the remainder to your answer   9
                                  + 1
                                  ———
                                   10

Find the number inside the division sign? You got it right.

9.1. Test Yourself

Do this division and check it:

$$2 \overline{)5}$$

Check your answer. If it is right, go to section 9.2. If it is wrong or you don't know what to do, go to section 9.1A.

9.1A. Ask yourself: How many times does 2 go into 5, or how many groups of 2 objects can I make with 5 objects? (If you need it, make a drawing.)

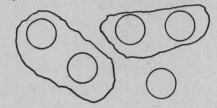

The answer is: 2 groups, with 1 object left over.
The answer to the division is: 2r1.
Check the division and go to 9.2.

9.2. Practice

Divide and check. Make sure your remainder is less than the number you divided by.

(a) 6) 8   (b) 4) 7   (c) 5) 17   (d) 6) 21   (e) 7) 39   **(f)** 43 ÷ 2 =
                                                                          (g) 37 ÷ 3 =
                                                                          (h) 65 ÷ 3 =

Check your answers.

10. REMAINDERS IN THE MIDDLE OF A DIVISION

When you divide a two-digit number, you may have a remainder in the middle of your division.
Example: 78 ÷ 3

Use the other division sign.
Divide the tens column. 3 goes into 7 two times. Put 2 in the tens column.

$$\begin{array}{r} \text{T U} \\ 2 \\ 3 \overline{)\, 7\ 8} \end{array}$$

But 2 × 3 is only 6. One is left over. JOIN this remainder with the units column. Now you have 18 units.

$$\begin{array}{r} 2\ \ 6 \\ 3 \overline{)\, 7\ {}^{1}8} \end{array}$$

Divide the units column. 3 goes into 18 six times. Put 6 in the units column.
Your answer: 26.

$$\begin{array}{r} \textcircled{1} \\ 2\ \ 6 \\ \times\ \ \ \ 3 \\ \hline 7\ \ 8 \end{array}$$

Check.

You may have a remainder in the middle of your division, and also at the end.
Example: 65 ÷ 4

Use the other division sign.
Divide the tens column. 4 goes into 6 one time. Put 1 in the tens column.

$$\begin{array}{r} 1 \\ 4\overline{)6\ 5} \end{array}$$

But 1 × 4 is only 4. 2 is left over. Join this remainder with the units column.

$$\begin{array}{r} 1 \\ 4\overline{)6\ ^25} \end{array}$$

Divide the units column. 4 goes into 25 6 times. Put 6 in the units column.

$$\begin{array}{r} 1\ 6 \\ 4\overline{)6\ ^25} \end{array}$$

But 4 × 6 is only 24. One is left over. There is nothing more to divide. Write r1 after the 6.

Your answer: 16r1

$$\begin{array}{r} 1\ 6\ \ r1 \\ 4\overline{)6\ ^25} \end{array}$$

Check.

$$\begin{array}{r} ②\ \\ 1\ 6 \\ \times\ \ \ 4 \\ \hline 6\ 4 \\ +\ \ \ \ 1\ \ \text{(remainder)} \\ \hline 6\ 5 \end{array}$$

10.1. Practice

(a) 4)76   **(b)** 5)77   **(c)** 8)92   **(d)** 3)59   **(e)** 99 ÷ 8 =

Check your answers. If they are all right, go to section 10.2. If at least one is wrong, read section 10 again very carefully and then go to section 10.2.

10.2. Sometimes the number in the last column is too small to be divided. This number becomes the remainder.
Example: 841 ÷ 4

Use the correct sign.
Divide the hundreds column.
Divide the tens column.

$$\begin{array}{r} 2\ 1 \\ 4\overline{)8\ 4\ 1} \end{array}$$

Divide the units column . . . but 4 doesn't go into 1. Put zero in units column (4 goes into 1 zero times).

$$\begin{array}{r} 2\ 1\ 0 \\ 4\overline{)8\ 4\ 1} \end{array}$$

4 × 0 is only 0. 1 is left over. There is nothing more to divide. Write r1 after the 0.
Your answer: 210r1.

$$\begin{array}{r} 2\ 1\ 0\ \ r1 \\ 4\overline{)8\ 4\ 1} \end{array}$$

Check.

$$\begin{array}{r} 2\ 1\ 0 \\ \times\ \ \ \ \ 4 \\ \hline 8\ 4\ 0 \\ +\ \ \ \ \ \ 1\ \ \text{(remainder)} \\ \hline 8\ 4\ 1 \end{array}$$

10.3. Practice

Divide and check:

(a) 5) 653　(b) 7) 846　(c) 4) 762　(d) 6) 543　(e) 3) 752

Check your answers.

11. DIVIDING LARGER NUMBERS

To divide a four-digit number, or a five-digit number, or any kind of number, follow the same steps as before.
Example:　$8,060 \div 6$

Use correct division sign.

Divide the thousands column. 6 goes into 8 one time, remainder 2. Put 1 in the thousands column, join 2 with the hundreds column.

$$\begin{array}{r} 1 \\ 6 \overline{\smash{)}8,\,^20\ \ 6\ \ 0} \end{array}$$

Divide the hundreds column. 6 goes into 20 three times, remainder 2. Put 3 in the hundreds column; join 2 with the tens column.

$$\begin{array}{r} 1\ \ 3 \\ 6 \overline{\smash{)}8,\,^20\ \ ^26\ \ 0} \end{array}$$

Divide the tens column. 6 goes into 26 four times, remainder 2. Put 4 in the tens column; join 2 to the units column.

$$\begin{array}{r} 1\ \ 3\ \ 4 \\ 6 \overline{\smash{)}8,\,^20\ \ ^26\ \ ^20} \end{array}$$

Divide the units column. 6 goes into 20 three times, remainder 2. Put 3 in the units column. There is nothing more to divide: put r2 after the 3.

$$\begin{array}{r} 1\ \ 3\ \ 4\ \ 3\ \ r2 \\ 6 \overline{\smash{)}8,\,^20\ \ ^26\ \ ^20} \end{array}$$

Your answer:　1,343r2.

Check.

$$\begin{array}{r} ②\ \ ②\ \ ① \\ 1,\ 3\ \ 4\ \ 3 \\ \times\ \ \ \ \ \ \ \ \ \ 6 \\ \hline 8,\ 0\ 5\ 8 \\ +\ \ \ \ \ \ \ \ \ 2\ (r) \\ \hline 8,\ 0\ 6\ 0 \end{array}$$

11.1. Practice

Divide and check:

(a) 7) 861　(b) 4) 756　(c) 3) 555　(d) 6)9,474

(e) $6,085 \div 5 =$ (f) $753 \div 6 =$ (g) $7,580 \div 4 =$ (h) $8,894 \div 8 =$

Check your answers.

11.2. More Practice

(a) $6 \overline{)36}$ (b) $7 \overline{)77}$ (c) $2 \overline{)684}$ (d) $3 \overline{)6,393}$ (e) $4 \overline{)80}$

(f) $8 \overline{)2,457}$ (g) $4 \overline{)9,652}$ (h) $3 \overline{)401}$ (i) $8 \overline{)659}$ (j) $5 \overline{)22,035}$

Check your answers.

12. WORD PROBLEMS—DIVISION

Finding out how many times one number is contained in another number is called DIVISION. Division word problems are easier to solve if you look for division word clues. Some division word clues are:

(a) EACH—The amount due is 270 dollars. There will be three payments. Each payment will be how many dollars?

(b) REMAINING—98 cans of soup are to go on 4 shelves. How many cans on each shelf? How many cans remaining?

(c) HOW MANY . . . IN . . . —How many weeks in 280 days?

(d) LEFT OVER—The baker made 729 rolls and put 6 rolls in each package. How many packages? How many rolls left over?

(e) DIVIDE—Four winners divided the $760 dollar prize money equally. How many dollars did each person receive?

(f) SHARE—Four people shared 848 miles of driving. What was each person's share?

(g) EQUALLY—249 acres of farmland were equally marked off for three family members. How many acres were in each person's lot?

12.1. Solutions

Each: $\dfrac{90}{3 \overline{)270}}$ Each payment will be $90.

Remaining: $\dfrac{24\ r\ 2}{4 \overline{)98}}$ There are 24 cans on each shelf, and 2 cans remaining.

How many . . . in . . . : $\dfrac{40}{7 \overline{)280}}$ There are 40 weeks (of 7 days) in 280 days.

Left over: 121 r 3
                  6) 729   There are 121 packages and 3 rolls left over.

Divide: 190
                 4) 760   Each person received $190.

Share: 212
                 4) 848   Each person's share of the driving was 212 miles.

Equally: 83
                 3) 249   Each person had an 83-acre lot.

12.2. Word Problems

1. Nine buses carried 441 passengers to the All Star Game. Each bus had the same number of people. How many passengers were on each bus?

   (1) 50   (2) 38   (3) 49   (4) 46
   (5) none of these

   1. 1 2 3 4 5

2. The babysitter had 13 slices of carrots to give four children for a snack. Each child got an equal share of carrots. How many carrots remained?

   (1) r2   (2) r1   (3) r0   (4) r3
   (5) none of these

   2. 1 2 3 4 5

3. 368 office workers work in an 8-story building. Each floor has the same number of people. How many workers are there on each floor?

   (1) 48   (2) 39   (3) 36   (4) 47
   (5) none of these

   3. 1 2 3 4 5

4. A wage earner paid the same amount of money for food and rent each month. The total came to 398 dollars. How much was each expense?

   (1) $189   (2) $196   (3) $216   (4) $199
   (5) none of these

   4. 1 2 3 4 5

5. A florist had 92 roses and put the same number of flowers into eight vases. How many roses went into each vase? How many roses were left over?

   (1) 11r4   (2) 12   (3) 14   (4) 11r5
   (5) none of these

   5. 1 2 3 4 5

6. 366 cans of soda were packed 6 cans to a carton. How many cartons of soda were there?

   (1) 51   (2) 16   (3) 36   (4) 61
   (5) none of these

   6. 1 2 3 4 5

7. 315 people work in a hospital on three 8-hour shifts. All shifts have the same number of people. How many people work each shift?

    7. 1 2 3 4 5

    (1) 115   **(2)** 150   **(3)** 105   **(4)** 15
    (5) none of these

8. You have 6 months to finish paying 186 dollars that you owe for a freezer. What are your six monthly payments?

    8. 1 2 3 4 5

    (1) $31   **(2)** $22   **(3)** $30   **(4)** $43
    (5) none of these

Check your answers. If all your answers are right, begin Chapter 9. If one or two are wrong, make sure that you understand the right answers and then start Chapter 9. If you made more than two mistakes, you should read Chapter 8 again before you start Chapter 9.

13. ANSWERS

1.1. 5

3.1. (a) 5   (b) 3   (c) 5   (d) 3   (e) 7

4.1. (a) 0   (b) 0   (c) impossible (d) impossible (e) impossible

      (f) 0   (g) 0   (h) 0   (i) impossible (j) impossible

5.2. 23

5.3. (a) 24   (b) 21   (c) 11   (d) 11   (e) 21

6.2. (a) 213   (b) 212   (c) 321   (d) 101   (e) 110   (f) 423

      (g) 121   (h) 101

7.1. (a) 3,122   (b) 2,112   (c) 2,312   (d) 2,121   (e) 1,110

7.3. 51

8.1. (a) 92   (b) 107   (c) 108   (d) 71   (e) 104   (f) 91

      (g) 205   (h) 41

9.1. 2r1

9.2. (a) 1r2   (b) 1r3   (c) 3r2   (d) 3r3   (e) 5r4   (f) 21r1

      (g) 12r1   (h) 21r2

10.1. (a) 19   (b) 15r2   (c) 11r4   (d) 19r2   (e) 12r3

10.3. (a) 130r3   (b) 120r6   (c) 190r2   (d) 90r3   (e) 250r2

11.1. (a) 123　　**(b)** 189　　**(c)** 185　　**(d)** 1,579　　**(e)** 1,217　　**(f)** 125r3

　　　　(g) 1,895　　**(h)** 1,111r6

11.2. (a) 6　　**(b)** 11　　**(c)** 342　　**(d)** 2,131　　**(e)** 20

　　　　(f) 307r1　　**(g)** 2,413　　**(h)** 133r2　　**(i)** 82r3　　**(j)** 4,407

12.1 Word Problems

1. (3)
$$
\begin{array}{r}
49 \\
9\,\overline{)\,441} \\
-36 \\
\hline
81 \\
-81 \\
\hline
\end{array}
$$

2. (2)
$$
\begin{array}{r}
3r1 \\
4\,\overline{)\,13} \\
-12 \\
\hline
1
\end{array}
$$

3. (5)
$$
\begin{array}{r}
46 \\
8\,\overline{)\,368} \\
-32 \\
\hline
48 \\
-48 \\
\hline
\end{array}
$$

4. (4)
$$
\begin{array}{r}
\$199 \\
2\,\overline{)\$398}
\end{array}
$$

5. (1)
$$
\begin{array}{r}
11r4 \\
8\,\overline{)\,92} \\
-8 \\
\hline
12 \\
-8 \\
\hline
4
\end{array}
$$

6. (4)
$$
\begin{array}{r}
61 \\
6\,\overline{)\,366} \\
-36 \\
\hline
6 \\
-6 \\
\hline
\end{array}
$$

7. (3)
$$
\begin{array}{r}
105 \\
3\,\overline{)\,315}
\end{array}
$$

8. (1)
$$
\begin{array}{r}
\$31 \\
6\quad \$186 \\
-18 \\
\hline
6 \\
-6 \\
\hline
\end{array}
$$

NINE

Division: Part II

1. DIVIDING BY A TWO-DIGIT NUMBER

Example: 84 ÷ 28
Put the division in usual form.
You need to find how many times 28 goes into 84.

```
    T U
28 )8  4
```

Try 4.
But 28 × 4 = 112; so 4 is too large.

```
        4
28 )8   4
    1 1 2
```

Try 3.
28 × 3 = 84. 3 is the right answer. Write it in the units column.
Subtract your answer to see if there is a remainder. There is none.

```
        3
28 )8   4
  - 8   4
    0   0
```

Your answer: 3.

Check by multiplying.

```
  ②
  2  8
×    3
  8  4
```

When you guess, you don't always pick the right number at once. You will know you have picked the wrong number if:
  —when you multiply, you find a number larger than the number inside the division sign;
or if:
  —when you find the remainder, it is not smaller than the number you divided by.
Here is an example of two wrong guesses:

First guess: 6
But 15 × 6 = 90; so 6 is too large.

```
        6
15 )7   5
  - 9   0
```

Second guess: 3
But 15 × 3 = 45, and the remainder is 30 (more than 15); so 3 is too small.

```
        3
15 )7   5
  - 4   5
    3   0
```

95

Third guess: 5

$15 \times 5 = 75$. 5 is the right answer. Write it in the units column.

Subtract your answer to see if there is a remainder.

There is none.

Your answer is 5.

Check.

```
          T U
            5
     15 ) 7 5
        - 7 5
        ─────
          0 0
   ──────────────
            ② 5
            1
          ×   5
          ─────
            7 5
```

1.1. Test Yourself

Do this division and check it by multiplying:

$$19 \overline{) 95}$$

Check your answer. If it is right, go to section 1.2. If it is wrong or you don't know what to do, go to section 1.1A.

1.1A. You must find how many times 19 goes into 95.

Try a few numbers.

If you try 7, you'll find that it is too large: $19 \times 7 = 133$.

So try a smaller number.

If you try 4, you'll see that it is too small: $19 \times 4 = 76$, and the remainder would be 19.

Try other numbers between 4 and 7 until you get the right answer.

Then check by multiplying.

Check your answer and go to section 1.2.

1.2. Practice

Divide and check. Use the correct division sign.

(a) $22 \overline{) 88}$ **(b)** $17 \overline{) 68}$ **(c)** $32 \overline{) 96}$ **(d)** $68 \div 34 =$ **(e)** $98 \div 14 =$

Check your answers.

1.3. If the number you are dividing by doesn't go exactly into the number inside the $\overline{)}$ sign, you'll have a remainder.

Remainders are no problem in division. Just remember: A REMAINDER MUST BE SMALLER THAN THE NUMBER YOU DIVIDE BY.

Example: $67 \div 14 =$

Put the division in usual form. You need to find how many times 14 goes into 67.

```
   T U
14 )6 7
```

Try 5. But $14 \times 5 = 70$, so 5 is too large.

```
        5
14 )6 7
  - 7 0
```

Try 4. $14 \times 4 = 56$.
Write 4 in the units column.
Subtract 56 to see if there is a remainder. The remainder is 11.
IT IS SMALLER THAN 14.
Your answer: 4r11.

```
        4  r 11
14 )6 7
  - 5 6
     1 1
```

Check.

```
      4
    × 4
    5 6
  + 1 1 (r)
    6 7
```

1.4. Practice

Divide and check.

      (a) 46)‾98 **(b)** 21)‾45 **(c)** 11)‾59 **(d)** $98 \div 32 =$ **(e)** $58 \div 27 =$

Check your answers.

2. DIVIDING A THREE-DIGIT NUMBER

Example: $222 \div 53 =$
Put the division in usual form.

```
    H T U
53 )2 2 2
```

Join the first two columns.
But 53 doesn't go into 22.

```
53 )2 2 2
```

Join the three columns.
How many times does 53 go into 222?
Guess: 5?
Multiply: $53 \times 5 = 265$. Wrong guess.
5 is too big.

```
        5
53 )2 2 2
  - 2 6 5
```

Guess: 4?

$53 \times 4 = 212$. The remainder is 10.

Write 4 in the units column.

Your answer: 4r10.

```
          4  r 10
53 ) 2  2  2
   - 2  1  2
          1  0
```

```
      ①
      5  3
   ×     4
   2  1  2
 +    1  0  (r)
   2  2  2
```

Check.

2.1. Practice

    (a) 48 $\overline{)\ 192}$ **(b)** 26 $\overline{)\ 228}$ **(c)** 17 $\overline{)\ 130}$ **(d)** $165 \div 23 =$ **(e)** $364 \div 91 =$

Check your answers.

3. DIVIDING MORE THAN ONCE

Sometimes you can start dividing after you have joined the first two columns. You must keep dividing until all the numbers have been used.

Example: $736 \div 23 =$

Put the division in usual form.

Join the first two columns: you have 73 tens.

23 goes into 73 three times. Put 3 in the tens column.

```
         3
23 ) 7  3  6
   - 6  9
         4
```

$23 \times 3 = 69$. The remainder is 4.

Bring the 6 down and join the last two columns: you have 46 units. 23 goes into 46 twice. Write 2 in the units column.

$23 \times 2 = 46$. Subtract your answer to see if there is a remainder. There is none.

```
         3  2
23 ) 7  3  6
   - 6  9
         4  6
       - 4  6
            0
```

```
      3  2
   ×  2  3
      9  6
    6  4  0
    7  3  6
```

Check.

Example: $390 \div 18 =$
Put the division in usual form.
Join the first two columns: you have 39 tens.
18 goes into 39 two times.
Put 2 in the tens column.
$18 \times 2 = 36$. The remainder is 3.

Bring the 0 down and join the last two columns: you have 30 units.
18 goes into 30 once. Write 1 in the units column.
$18 \times 1 = 18$. The remainder is 12.
There is nothing more to bring down.
Your answer: 21r12.

```
        2
  18 ) 3 9 0
      - 3 6
          3
```

```
        2 1  r 12
  18 ) 3 9 0
      - 3 6
          3 0
        - 1 8
          1 2
```

```
          2 1
      ×   1 8
        1 6 8
        2 1 0
        3 7 8
      +   1 2
        3 9 0
```

Check.

3.1. Test Yourself

Do this division and check it by multiplying:

```
  16 ) 572
```

Check your answer. If it is right, go to section 3.2. If it is wrong or you don't know what to do, go to section 3.1A.

3.1A. Here is how to start the division:

```
          3
  16 ) 5 7 2
      - 4 8
```

Now you can finish it. Check it by multiplying. (Don't forget to add the remainder.) Then go to section 3.2.

3.2. Practice

   (a) 26) 488   **(b)** 54) 657   **(c)** 17) 529   **(d)** 11) 576   **(e)** 15) 789

Check your answers.

4. ZERO IN THE QUOTIENT

Sometimes, while you are doing your division, you bump into a number that is too small to be divided.

Example: $759 \div 37 =$

Put the division in usual form. Join the first two columns: you have 75 tens.

37 goes into 75 two times. Write 2 in the tens column.

$37 \times 2 = 74$. The remainder is 1.

Bring the 9 down and join the columns: you have 19 units.

37 doesn't go into 19: it goes zero times.

Put 0 in the units column.

$37 \times 0 = 0$. Subtract your answer from 19 to find the remainder: the remainder is 19.

Your answer: 20r19.

```
    H  T  U
          2
37 ) 7  5  9
   - 7  4
          1
```

```
       2  0
37 ) 7  5  9
   - 7  4
       1  9
```

```
       2  0  r 19
37 ) 7  5  9
   - 7  4
       1  9
    -     0
       1  9
```

Check.

```
         2  0
      ×  3  7
      1  4  0
      6  0  0
      7  4  0
   +     1  9
      7  5  9
```

Make sure that you write the zero down. If you don't you will get a 2 as an answer rather than 20. (Would you rather have $2 or $20?)

4.1. Practice

    (a) 17) 689   **(b)** 37) 752   **(c)** 24) 975   **(d)** 43) 878   **(e)** 16) 493

Check your answers.

5. DIVIDING LARGER NUMBERS

To divide a four-digit number or a five-digit number, or any kind of number, follow the same steps.

Example: 3,579 ÷ 42 =

Put the division in usual form.

Join the first two columns. But 42 doesn't go into 35.

So join the first three columns: you have 357 tens.

42 goes into 357 eight times. Write 8 in the tens column.

42 × 8 = 336. Remainder: 21.

Bring down the 9 and join the columns: you have 219 units.

42 goes into 219 five times. Write 5 in the units column.

42 × 5 = 210. Remainder: 9.

There is nothing more to bring down.

Your answer: 85r9.

5.1. Practice

Divide and check.

(a) 26) 589   **(b)** 30) 748   **(c)** 62) 439   **(d)** 483 ÷ 19 = **(e)** 84,293 ÷ 71 =

Check your answers.

5.2. More Practice

(a) 36) 74   **(b)** 58) 587   **(c)** 35)3,694   **(d)** 29)4,578

(e) 65) 762   **(f)** 43)5,162   **(g)** 12)7,894   **(h)** 82) 410

Check your answers. If they are all right, start Chapter 10. If one or two are wrong, make sure that you understand the right answers and then start Chapter 10. If you made more than two mistakes, you should read Chapter 9 again before starting Chapter 10.

MATHEMATICS

6. ANSWERS

1.1. 5

1.2. (a) 4   **(b)** 4   **(c)** 3   **(d)** 2   **(e)** 7

1.4. (a) 2r6   **(b)** 2r3   **(c)** 5r4   **(d)** 3r2   **(e)** 2r4

2.1. (a) 4   **(b)** 8r20   **(c)** 7r11   **(d)** 7r4   **(e)** 4

3.1. 35r12

3.2. (a) 18r20   **(b)** 12r9   **(c)** 31r2   **(d)** 52r4   **(e)** 52r9

4.1. (a) 40r9   **(b)** 20r12   **(c)** 40r15   **(d)** 20r18   **(e)** 30r13

5.1. (a) 22r17   **(b)** 24r28   **(c)** 7r5   **(d)** 25r8   **(e)** 1,187r16

5.2. (a) 2r2   **(b)** 10r7   **(c)** 105r19   **(d)** 157r25   **(e)** 11r47

     (f) 120r2   **(g)** 657r10   **(h)** 5

TEN

Division: Part III

1. DIVIDING BY A THREE–DIGIT NUMBER

By now you have discovered why this kind of division is called LONG DIVISION! It does take some time, but by doing all the steps carefully, you will get correct answers.

To divide by a three-digit number follow the same steps.

Example: 969 ÷ 323
Put the division in usual form.
You are going to divide by 3 digits, so you need to join three columns.
You must find how many times 323 goes into 969. Guess until you find the right answer. (Remember the rules for guessing.)
323 goes into 969 three times. Write 3 in the units column.
323 × 3 = 969. There is no remainder.
There is nothing to bring down.
Your answer: 3.

Check.

```
      H T U
323 )9 6 9


            3
323 )9 6 9
   - 9 6 9
     0 0 0


     3 2 3
   ×     3
     9 6 9
```

Example: 680 ÷ 127
Put the division in usual form.
Join three columns.
You must find how many times 127 goes into 680. Guess until you find the right answer.

127 goes into 680 five times. Write 5 in the units column.
127 × 5 = 635. The remainder is 45. Write it with your answer.
There is nothing to bring down.
Your answer: 5r45.

Check.

```
127 )6 8 0


          5 r45
127 )6 8 0
   - 6 3 5
       4 5

     1 2 7
   ×     5
     6 3 5
   +   4 5
     6 8 0
```

103

1.1. Test Yourself

Do this division and check it:

    957 ÷ 239

Check your answer. If it is right, go to section 1.2. If it is wrong or if you don't know what to do, go to section 1.1A.

1.1A. Write your division using the other division sign. Join the three columns before you start:

    239) $\overline{957}$

You guess how many times 239 goes into 957.

9 is too big; 7 is too big.

Continue guessing until you find the right answer.

Check your division by multiplying. (Don't forget to add the remainder.) Then go on to section 1.2.

1.2. Practice

Divide and check.

    (a) 136) $\overline{952}$ **(b)** 689) $\overline{698}$ **(c)** 248) $\overline{997}$ **(d)** 352) $\overline{984}$ **(e)** 186) $\overline{759}$

Check your answers.

2. DIVIDING LARGER NUMBERS

When you divide a four-digit number or a five-digit number, or any kind of number, you may have to divide more than once. Always follow the same steps.

Example: 64,959 ÷ 468

Put the division in usual form. Join first three columns.

468 goes into 649 once. Write 1 in hundreds column.

468 × 1 = 468. Remainder: 181.

TTH	TH	H	T	U
		1		
468) 6	4,	9	5	9
	– 4	6	8	
	1	8	1	

Bring the 5 down and join the columns: you have 1,815 tens.

468 goes into 1,815 three times. Put 3 in the tens column.

468 × 3 = 1,404. Remainder: 411.

TTH	TH	H	T	U
		1	3	
468) 6	4,	9	5	9
	– 4	6	8	
	1	8	1	5
	– 1	4	0	4
		4	1	1

Bring the 9 down and join the columns: you have 4,119 units.
468 goes into 4,119 eight times. Put 8 in the units column.
468 × 8 = 3,744. Remainder: 375.
There is nothing more to bring down.
Your answer: 138r375.   (375 is smaller than 468.)

```
          TTH TH  H  T  U
                  1  3  8r375
     468 )6  4,  9  5  9
         -4   6  8
          1   8  1  5
         -1   4  0  4
              4  1  1  9
             -3  7  4  4
                 3  7  5
```

Check.

```
                4  6  8
          ×        1  3  8
                3  7  4  4
             1  4  0  4  0
             4  6  8  0  0
             6  4  5  8  4
          +        3  7  5
             6  4,  9  5  9
```

2.1. Practice

(a) 324)74,389   (b) 648)1,452   (c) 542)26,215   (d) 843 ÷ 729 =

(e) 8,548 ÷ 169 =

Check your answers.

3. DIVIDING BY 10, 100, 1,000: SHORTCUTS

To divide by 10, cross out the last zero.

```
              497
     10 )4,970
         -4 0
            97
          - 90
             70
          -  70
```

Long way

```
            4  9  7
     10 ) 4, 9  7  0
```
Short way

To divide by 100, cross out the last two zeros.

$$\begin{array}{r} 8,327 \\ \hline 100 \enclose{longdiv}{832,700} \end{array}$$

Short way

To divide by 1,000, cross out the last zeros.

$$\begin{array}{r} 3,578 \\ \hline 1,000 \enclose{longdiv}{3,578,000} \end{array}$$

Short way

The short way saves time and space.

3.2. Practice

(a) $10 \enclose{longdiv}{7,640}$ **(b)** $100 \enclose{longdiv}{54,500}$ **(c)** $1,000 \enclose{longdiv}{6,324,000}$

(d) $9,430 \div 10 =$ **(e)** $26,200 \div 100 =$

Check your answers.

4. YOU CAN DIVIDE ANY NUMBERS

In Chapters 8, 9 and 10, you have been learning how to divide ANY NUMBERS. No matter how many digits they have, no matter how many times you have to divide, just remember the rules you have learned so far.

5. DIVISION REMINDERS

If division problems are in this form:

$$256 \div 23$$

be sure to put them in this form:

$$23 \enclose{longdiv}{256}$$

before you start doing them.
Divide FROM LEFT TO RIGHT.

When you divide by a two-digit number, join the first two columns.
When you divide by a three-digit number, join the first three columns.
You may have to join more columns before you can start.
Start writing your answers in the last column you joined:

$$\begin{array}{r} 1 \\ 23 \overline{)2\ \ 5\ \ 6} \end{array} \qquad \begin{array}{r} 7 \\ 62 \overline{)4\ \ 8\ \ 9\ \ 5} \end{array}$$

Then you must have a digit all the way to the right:

$$\begin{array}{r} 1\ \ *\ \ *\ \ * \\ 34 \overline{)6\ \ 5,\ 9\ \ 4\ \ 3} \end{array} \quad \begin{array}{l} \text{first answer} \\ \text{must have digits in each place} \end{array}$$

Always put your answers in the correct columns.
Do not forget to write the zeros.
Show all your work.
REMAINDERS MUST BE LESS THAN THE NUMBER YOU DIVIDE BY.
Give your remainder together with your answer.
Check your answer by multiplying. Add the remainder if there is one.
Remember the short way to divide by 10, 100, 1,000.

5.1. More Practice

(a) $361 \overline{)1,444}$ (b) $750 \overline{)4,678}$ (c) $10 \overline{)6,280}$ (d) $513 \overline{)4,134}$

(e) $222 \overline{)693,489}$ (f) $438 \overline{)26,958}$ (g) $100 \overline{)329,600}$

Check your answers.

5.2. Word Problems

Before you start, go back to Chapter 8 to remind yourself of the DIVISION WORD CLUES.

1. A hardware store bought 2,575 feet of rope and sold it in packages of 25 feet. How many packages were made from the rope?
   (1) 1,105   (2) 103   (3) 13   (4) 10r3
   (5) none of these

1. 1 2 3 4 5

2. There are 140 bricks in a bundle. If there are 3,284 bricks. How many bundles will there be? How many loose bricks will remain?

2. 1 2 3 4 5

(1) 23   (2) 234   (3) 264   (4) 23r64

(5) none of these

3. You need 38 inches of 2-foot wide storm plastic to cover a window. How many windows could be covered with 456 inches of plastic?

3. 1 2 3 4 5

(1) 12   (2) 19   (3) 228   (4) 14

(5) none of these

4. If 10 equal sized water tanks hold a total of 13,640 gallons of water, how many gallons of water does each tank hold?

4. 1 2 3 4 5

(1) 136r4   (2) 13r64   (3) 1,364   (4) 136

(5) none of these

5. You borrow $1,200 from the bank and agree to pay it back in 24 equal monthly payments. What is your payment each month?

5. 1 2 3 4 5

(1) $24   (2) $55   (3) $20   (4) $50

(5) none of these

6. 236 produce trucks carry 968,537 pounds of produce across the nation. Each truck carries the same weight. How many pounds of produce does each truck carry?

6. 1 2 3 4 5

(1) 4033r50   (2) 4103r129   (3) 34,103   (4) 4,100

(5) none of these

7. Grace's car gets about 19 miles to a gallon of gasoline. How many gallons does the car use to travel 1,900 miles?

7. 1 2 3 4 5

(1) 19   (2) 1,000   (3) 100   (4) 10

(5) none of these

8. A factory can produce 200 toasters in a day. How many days would it take to make 18,600 toasters?

8. 1 2 3 4 5

(1) 9   (2) 93   (3) 93r10   (4) 930

(5) none of these

Check your answers. If all your answers are right, go on to Evaluation: Four Operations. If one or two answers are wrong, make sure that you understand the right answers and then start the evaluation. If more than two answers are wrong, you should reread Chapter 10 before starting the evaluation.

6. ANSWERS

1.1.
```
           4r1
239 ) 957
      956
      ----
        1
```

1.2. (a) 7   **(b)** 1r9   **(c)** 4r5   **(d)** 2r280   **(e)** 4r15

2.1. (a) 229r193   **(b)** 2r156   **(c)** 48r199   **(d)** 1r114   **(e)** 50r98

3.2. (a) 764   **(b)** 545   **(c)** 6,324   **(d)** 943   **(e)** 262

5.1. (a) 4   **(b)** 6r178   **(c)** 628   **(d)** 8r30   **(e)** 3,123r183

         (f) 61r240   **(g)** 3,296

5.2. Word Problems

1. (2)
```
          103
25 )2,575
   - 25
     ----
      75
    - 75
     ----
```

2. (4)
```
           23r64
140 ) 3,284
    - 2 80
      -----
        484
      - 420
      -----
         64
```

3. (1)
```
          12
38 ) 456
   - 38
     ----
      76
    - 76
     ----
```

4. (3)
```
         1,364
10 ) 13,640
```

5. (4)
```
         $  50
24 )$1,200
   - 1 20
     ------
        0
```

6. (5)
```
          4,103r229
236 ) 968,537
    - 944
      -----
       24 5
     - 23 6
      -----
         937
       - 708
       -----
         229
```

7. (3)
```
         100
19 )1,900
```

8. (2)
```
          93
200 )18,600
```

Evaluation: Four Operations

This is to find out what you know and what you still need to work on. Do each problem carefully. Take your time and watch the signs.

NUMBER PROBLEMS

1. 532
   × 25

2. 2,582
   + 6,957

3. 6)9,534

4. 1,043
   − 652

5. 23
   + 36

6. 304
   201
   + 39

7. 1,001
   − 640

8. 508)25,943

9. 413
   × 3

10. 901
    − 430

11. 57
    + 35

12. 2,950
    3,087
    + 5,392

13. 87
    + 64

14. 584
    − 396

15. 26) 234

16. 2,563
    − 1,092

17. 4,382
    × 1,000

18. 564,259
    + 201,787

19. 8,206
    − 4,319

20. 257
    × 18

21. 400,752
    + 699,398

22. 36,782
    − 19,435

23. 615
    × 203

24. 5)96,585

25. 957
    × 382

26. 324)986,135 27. 92)98,593 28.   497   29. 10)5,790 30.   78
                                         − 263   − 32

31. 81)3,245 32.   34   33.   908   34. 100)982,400 35.   131
                   × 2   × 5   97
                                                             + 26

36. 17)3,520 37. 4,396 38.   216   39.   257   40.   589
                  × 208   × 34   − 172   1,024
                                                         + 3,876

WORD PROBLEMS

41. A drug store had 24 electric razors and 54 hand razors in stock. How many razors did the store have in all?
  (1) 77   (2) 79   (3) 68   (4) 78
  (5) none of these

41. 1 2 3 4 5

42. The population of Arkansas was 2,000,000 in 1965. Over the next ten years, the population increased by 117,000. What was the population of Arkansas in 1975?
  (1) 2,100,010   (2) 2,117,000   (3) 2,011,700
  (4) 2,000,117   (5) none of these

42. 1 2 3 4 5

43. A high school had 458 freshmen, 387 sophomores, 396 juniors, and 408 seniors. How many students were enrolled in this high school?
  (1) 1,759   (2) 1,649   (3) 1,638   (4) 1,369
  (5) none of these

43. 1 2 3 4 5

44. When Sy bought his car, there were already, 12,784 miles on it. A year later the odometer said there were 16,997 miles on it. How many miles did Sy drive the car in the year?
  (1) 4,213   (2) 4,173   (3) 3,273   (4) 4,611
  (5) none of these

44. 1 2 3 4 5

45. The baker made 48 sweet rolls and sold all but 6. How many rolls did she sell?
  (1) 54   (2) 44   (3) 34   (4) 42
  (5) none of these

45. 1 2 3 4 5

46. In the last election, there were 6,349 registered to vote. If 2,873 people did not vote, how many people did vote?
    (1) 3,476   (2) 3,572   (3) 4,476   (4) 3,572
    (5) none of these

46. 1 2 3 4 5

47. Roberto bought a stereo system for $438. After three years he sold it for $245. How much did he reduce the price of the stereo from its original price?
    (1) $228   (2) $195   (3) $193   (4) $173
    (5) none of these

47. 1 2 3 4 5

48. The number of people who received public assistance benefits in 1975 was 5,019. In 1977, this figure was reduced by 1,395. How many people received benefits in 1977?
    (1) 3,605   (2) 3,624   (3) 4,604   (4) 3,734
    (5) none of these

48. 1 2 3 4 5

49. A storage tank holds 150,000 gallons of water. Three tanks of the same size are under construction. How much water will the new tanks hold altogether?
    (1) 153,000   (2) 40,000   (3) 450,000   (4) 45,000
    (5) none of these

49. 1 2 3 4 5

50. A factory manufactures 12,500 hubcaps a day. How many hubcaps are manufactured in a five-day work week?
    (1) 6,250   (2) 625,000   (3) 62,500   (4) 6,520
    (5) none of these

50. 1 2 3 4 5

Check your answers. If you made any mistakes, make sure that you understand the right answers. Go back to any chapters that you need to review again.
Then go on to Chapter 11.

ANSWERS

1. 13,300	2. 9,539	3. 1,589	4. 391	5. 59
6. 544	7. 361	8. 51r35	9. 1,239	10. 471
11. 92	12. 11,429	13. 151	14. 188	15. 9
16. 1,471	17. 4,382,000	18. 766,046	19. 3,887	20. 4,626
21. 1,100,150	22. 17,347	23. 124,845	24. 19,317	25. 365,574
26. 3,043r203	27. 1,071r61	28. 234	29. 579	30. 46
31. 40r5	32. 68	33. 4,540	34. 9,824	35. 254
36. 207r1	37. 914,368	38. 7,344	39. 85	40. 5,489

41. (4) 24
$+$ 54
___
78

42. (2) 2,000,000
$+$ 117,000
___
2,117,000

43. (2) 458
387
396
$+$ 408
___
1,649

44. (1) 16,997
$-$ 12,784
___
4,213

45. (4) 48
$-$ 6
___
42

46. (1) 6,349
$-$ 2,873
___
3,476

47. (3) $438
$-$ $245
___
$193

48. (2) 5,019
$-$ 1,395
___
3,624

49. (3) 150,000
$\times$ 3
___
450,000

50. (3) 12,500
$\times$ 5
___
62,500

ELEVEN

Fractions

1. WHAT IS A FRACTION?

A *fraction* is a *part* of a unit or a part of a group.

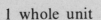

1 whole unit   2 equal parts

1 whole group   3 equal parts

1 whole unit   2 equal parts

1 whole group   2 equal parts

2. NAMES OF FRACTIONS

2 equal parts
Each part is called
a half

4 equal parts
Each part is called
a fourth

3 equal parts
Each part is called
a third

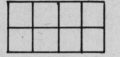

8 equal parts
Each part is called
an eighth

6 equal parts
Each part is called
a sixth

12 equal parts
Each part is called
a twelfth

It is easy to find the name of any fraction: if a unit or a group is divided into seven parts, each part is called a sevenTH. If a unit or group is divided into nine parts, each part is called a ninTH. Just take the number of parts that there are and add TH.

3. HOW TO WRITE FRACTIONS

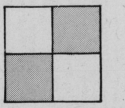

Number of dark parts $\dfrac{2}{4}$ (2 out of 4 are dark).

Number of parts in
whole square

The number on top of the bar is the NUMERATOR of the fraction; the number at the bottom is the DENOMINATOR of the fraction.

$\dfrac{2}{4}$ numerator   how many parts you're talking about

     denominator   how many parts in the whole thing

There are two ways to write a fraction. You can write it either this way: $^{2}/_{4}$ or this way: 2/4. The numerator is on top of the bar, and the denominator is at the bottom.

| Number of dark parts | $\dfrac{1}{2}$ | (1 out of 2 is dark) |
| Number of parts in whole circle | | |

| Number of dark parts | $\dfrac{1}{3}$ | (1 out of 3 is dark) |
| Total number of parts | | |

| Number of dark stars | $\dfrac{3}{5}$ | (3 out of 5 are dark) |
| Total number of stars in whole group | | |

| Number of dark parts | $\dfrac{3}{4}$ | (3 out of 4 are dark) |
| Total number of parts | | |

3.1. Test Yourself

(a) Write a fraction for the dark parts.

(b) Write a fraction for the bowling pins that are down.

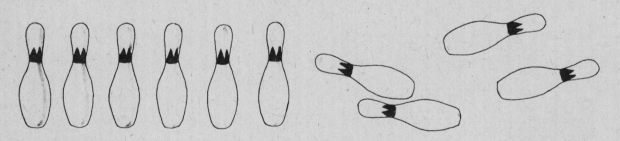

Check your answers. If both are right, go on to section 3.2. If one is wrong, go to section 3.1A.

3.1A. To write a fraction for the dark equal parts of the square, here is all you do:

How many parts are dark? ^{4}/

How many parts in the whole thing?   /₆

Name of the fraction: ^{4}/₆

To write a fraction for the number of bowling pins knocked down:

How many bowling pins are down? $^4/$

How many bowling pins in all? $/_{10}$

Name of the fraction: $^4/_{10}$

3.2. Practice

Write a fraction for the dark parts.

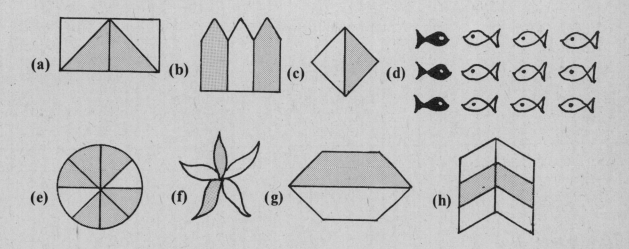

Check your answers. If they are not all right, read sections 3 and 3.1A again and then go on to section 3.3.

3.3. When a fraction has a numerator that is smaller than the denominator, it is called a PROPER fraction.

A proper fraction is less than something that is whole.

$$\frac{1}{4} \quad \frac{\text{numerator smaller than}}{\text{denominator}}$$

3.4. When a fraction has a numerator that is as large as its denominator, or a numerator that is larger than its denominator, it is called an IMPROPER fraction.

An improper fraction is something whole or more than something whole.

$$\frac{6}{4} \quad \frac{\text{numerator larger than}}{\text{denominator}}$$

3.5. Practice

For each fraction, decide if it is a proper fraction or an improper fraction.

(a) $\frac{3}{9}$ **(b)** $\frac{12}{8}$ **(c)** $\frac{4}{3}$ **(d)** $\frac{9}{7}$ **(e)** $\frac{1}{4}$

Check your answers.

4. COMPARING FRACTIONS

The fraction graphs give a picture of fractions. Use fraction graphs to compare fractions.

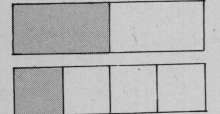

This graph is a picture of ½

This graph is a picture of ¼

The graphs show that ½ is larger than ¼.

4.1. Practice

Find the larger fraction

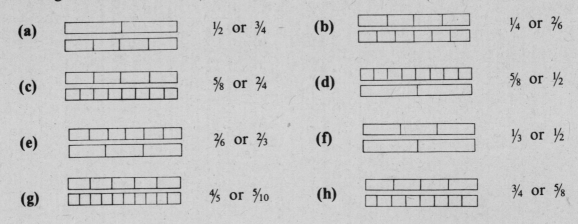

(a) ½ or ¾   **(b)** ¼ or ²⁄₆

(c) ⅝ or ²⁄₄   **(d)** ⅝ or ½

(e) ²⁄₆ or ⅔   **(f)** ⅓ or ½

(g) ⅘ or ⁵⁄₁₀   **(h)** ¾ or ⅝

Check your answers.

5. EQUIVALENT FRACTIONS

When you compare fractions by looking at their graphs, you may find that two fractions take up the same space.

Fractions with different names are equivalent (equal) if they take up the same space.

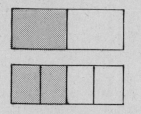

Dark part: ½

Dark part: ¾

Compare

Same amount of dark space?

They are equivalent.

You write:

½ = ¾ or
1/2 = 2/4

5.1. Test Yourself

Look at the graphs and write the equivalent fractions.

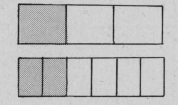

Check your answer. If it is right, go to section 5.2. If it is wrong or you don't know what to do, go to section 5.1A.

5.1A.

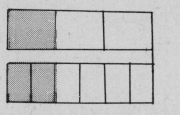

Number of dark parts:	1
Total number of parts:	3

Number of dark parts:	2
Total number of parts:	6

They take up the same space: so ⅓ = 2/6.

5.2. Practice

Look at the fraction graphs. Find the equivalent fractions.

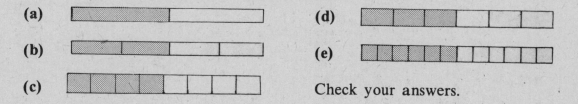

(a)

(d)

(b)

(e)

(c)

Check your answers.

6. MAKING FRACTION GRAPHS

Let us make fraction graphs to find the missing number: $\frac{1}{2} = \frac{?}{6}$

Make two graphs of the same size:

Divide the first graph in 2 parts and darken 1 part:

Divide the other in 6 parts.

How many parts must be dark so that you have the same amount of dark in both?

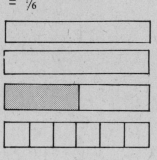

3 parts. The answer is $\frac{3}{6}$.

6.1. Practice

Make fraction graphs to find the missing numbers.

(a) $\frac{1}{2} = \frac{}{4}$ **(b)** $\frac{2}{4} = \frac{}{8}$ **(c)** $\frac{1}{3} = \frac{}{6}$ **(d)** $\frac{3}{6} = \frac{}{2}$ **(e)** $\frac{2}{4} = \frac{}{12}$

7. REDUCING TO LOWEST TERMS

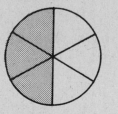

$\frac{3}{6}$ of the circle is dark,

or

$\frac{1}{2}$ of the circle is dark.

Reducing to lowest terms means finding the smallest possible number for the numerator and the denominator. The fraction still has the same value.

$\frac{4}{6}$
$\frac{2}{3}$ same space: $\frac{4}{6} = \frac{2}{3}$

$\frac{3}{9}$
$\frac{1}{3}$ same space: $\frac{3}{9} = \frac{1}{3}$

Here is what to do to reduce a fraction to lowest terms.
Example: reduce $\frac{4}{8}$

What numbers can divide 4 with no remainder?
1, 2, 4

What numbers can divide 8 with no remainder?
1, 2, 4, 8

What is the largest number that divides both with no remainder?
1, 2, 4
1, 2, 4, 8   It is 4.

$$\frac{4 \div 4}{8 \div 4} = \frac{1}{2}$$

$$\frac{4}{8} = \frac{1}{2}$$

Check with fraction graphs if you want to make sure.

7.1. Test Yourself

Reduce to lowest terms: $6/8$
Check your answer. If it is right, go to section 7.2. If it is wrong or you don't know what to do, go to section 7.1A.

7.1A. Numbers that can divide 6:   1, 2, 3, 6
Numbers that can divide 8:   1, 2, 4, 8
Find the largest number that divides both.
Then divide the numerator and the denominator of your fraction by this number.
Check your answer and go to section 7.2.

7.1B. Reducing fractions to lowest terms does not change the value.
A fraction is already in lowest terms when there is no number (except 1) that can divide both the numerator and the denominator.
Example: $2/3$
　　　Numbers that can divide 2: 1, 2
　　　Numbers that can divide 3: 1, 3
There is no number that divides both 2 and 3: this fraction is already in lowest terms.
Fraction answers must always be given in lowest terms.

7.2. Practice

Reduce these fractions. Some are already in lowest terms. You can check by making fraction graphs. You may also have to reduce improper fractions.

(a) $5/15$ **(b)** $10/100$ **(c)** $12/24$ **(d)** $12/60$ **(e)** $6/6$ **(f)** $3/9$ **(g)** $7/14$ **(h)** $9/12$

Check your answers.

8. RAISING TO HIGHER TERMS

Instead of reducing a fraction to lowest terms, you may have to raise it to higher terms.

Just multiply the numerator and the denominator by the same number.

Example: raise ⅔ to higher terms.

Pick a number such as 5, for example.

Multiply

$$\frac{2 \times 5}{3 \times 5} = \frac{10}{15}$$

The numerator and the denominator are now higher, but the amount is the same.

Sometimes you already know the denominator of the fraction with higher terms. So you only have to find the numerator.

There are 12 crew members on a boat. ⅔ of them have sailed many times. How many out of 12 have sailed many times?

Think: 3 × ? = 12

Answer: 4

Multiply the numerator and the denominator by 4.

So 8 out of 12 sailors have sailed many times.

8.1. Test Yourself

There are 12 eggs in a dozen. ⅓ of the dozen has already been used. How many eggs out of 12 have been used?

Check your answer. If it is right, go to section 9. If it is wrong or you don't know what to do, go to section 8.1A.

8.1A. Your problem is: ⅓ = ?/12

Think: 3 × ? = 12.

When you get the answer, multiply the numerator and the denominator of your fraction by this number.

Check your answer and go to section 9.

9. CHANGING WHOLE NUMBERS TO FRACTIONS

Whole numbers can be written as fractions. The fraction is improper, because the numerator is the same as, or larger than, the denominator.

one whole pie, uncut:

Number of dark parts	1
Total number of parts	

two whole loafs, uncut:

Number of dark parts	2
Total number of parts *in each loaf*	1

three whole melons:

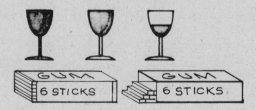

Number of dark parts	3
Total number of parts *in each melon*	1

To change a whole number to a fraction:

$$\frac{\text{put whole number}}{1}$$

9.2. Practice

Change these whole numbers to improper fractions.

(a) $5 = \frac{5}{1}$ (b) $6 = /$ (c) $8 = /$ (d) $12 = /$ (e) $13 = /$

Check your answers.

10. MIXED NUMBERS

A mixed number is a whole number plus a fraction.

2½ glasses of wine

1⅚ packs of gum

11. CHANGING IMPROPER FRACTIONS TO MIXED NUMBERS

Example: change ⁵⁄₄ to a mixed number.

Divide the numerator by the denominator.

$$4\overline{)\,5}$$

The quotient is your whole number.

$$\overset{1\ r1}{4\,)\,5}\qquad 1$$

The remainder and the denominator of the fraction you had before make the fraction.
Your answer is 1¼.

$$\overset{1\ r1}{4\,)\,5}\qquad 1¼$$

Important: the improper fraction you started with was ⁵⁄₄.

the mixed number you found is 1¼.

Always check that the two denominators are the same.

11.1. Test Yourself

Change ⁷⁄₃ to a mixed number.
Check your answer. If it is right, go to section 11.2. If it is wrong or you don't know what to do, go to section 11.1A.
11.1A. Remember how to divide.
Here is what your division should give you:

$$\overset{2r1}{3\,)\,7}$$

If you still have trouble with division, read Chapter 4 again.
11.2. Sometimes, after finding the mixed number, you still have to reduce its fraction to lowest terms.
Example: change ¹⁰⁄₄ to a mixed number

Divide the numerator by the denominator.

$$\overset{2\,r\,2}{4\,)\,10}\qquad 2$$

The quotient is your whole number, the remainder and the denominator of the fraction you had before make the fraction.

$$\overset{2\,r\,2}{4\,)\,10}\qquad 2¼$$

Your answer: 2¼

$$¾ = ½$$

You must reduce ¼ to lowest terms

$$2¾ = 2½$$

Your final answer: 2½

This time the denominator in your mixed number is not the same as in the fraction you started with because you reduced to lowest terms.

Reduce your answers to lowest terms every time you can.

11.3. Sometimes when you do your division you find no remainder. Then what you find is not a mixed number, but a whole number.

Example: change ¹⁸⁄₃ to a mixed number.

Divide the numerator by the denominator.

$3 \overline{)\ 18}$

The quotient is your whole number, and there is no remainder: so your answer is not a mixed number, but a whole number: 6.

$$3 \overline{)\ 18} \quad \overset{(6)}{} \quad 6$$

11.4. Practice

    (a) ⁸⁄₃   **(b)** ¹⁵⁄₃   **(c)** ¹⁴⁄₄   **(d)** ¹⁶⁄₇   **(e)** ¹⁸⁄₉

Check your answers.

12. CHANGING MIXED NUMBERS TO IMPROPER FRACTIONS

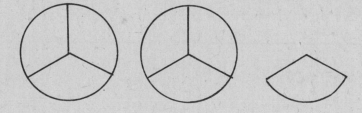

There are 2⅓ pies.
If we count the pieces, we find ⁷⁄₃.
So 2⅓ = ⁷⁄₃

Here is how to change a mixed number to an improper fraction.   2⅓

Multiply the whole number and the denominator.   2 × 3 = 6

Add the numerator.   6 + 1 = 7

Put the answer on top of the denominator.   ⁷⁄₃

Your answer is ⁷⁄₃.

Important: the mixed number you started with was   2⅓

               the improper fraction you found is   ⁷⁄₃

Always check that the two denominators are the same.

12.1. Test Yourself

Change 3½ to an improper fraction.

Check your answer. If it is right, go to section 12.2. If it is wrong or you don't know what to do, go to section 12.1A.

12.1A. You must multiply the whole number and the denominator, so you must find 3 × 2.

Add the numerator (1) to your answer.

Put the answer on top of the denominator. Write your fraction and go to section 12.2.

12.2. Practice

Change these mixed numbers to improper fractions.

    (a) 4½   **(b)** 3¼   **(c)** 5⅗   **(d)** 6⅞   **(e)** 2⁹⁄₁₀

    (f) 3⅔   **(g)** 2⁶⁄₉   **(h)** 3⅘

Check your answers.

13. FRACTION REMINDERS

A FRACTION is a part of a unit, or a part of a group.

The NUMERATOR is written on top of the fraction bar; the DENOMINATOR is at the bottom.

$$\frac{2}{5} \quad \text{numerator} \atop \text{denominator} \quad \text{or} \quad 2/5 \quad \text{numerator} \atop \text{denominator}$$

A PROPER fraction has a numerator that is smaller than its denominator; otherwise it is an IMPROPER fraction.

A fraction can be reduced to lowest terms:

    ⁴⁄₆ = ⅔ (these two fractions are equivalent: they have the same value)

A fraction can be raised to higher terms:

    ⅖ = ⁸⁄₂₀ (these two fractions are equivalent)

A whole number can be written as a fraction:

    7 = ⁷⁄₁ (the denominator is always 1)

A mixed number is a whole number plus a fraction:

    3¼

A mixed number can be changed to an improper fraction, and an improper fraction can be changed to a mixed number.

14. WORD PROBLEMS—FRACTIONS

FRACTIONS are parts of a whole or parts of a group. Fraction word problems are easier to solve if you look for fraction word clues. Some fraction word clues are:

(a) PART—What is the fraction name for each part of a parking lot roped off into 8 equal sections?

(b) SHARE—4 investors marked off a building lot into 4 equal lots. What was each investor's share?

(c) RATIO—(a) 6 trucks and 11 cars parked in a garage. What is the ratio of trucks to cars?

   (b) 8 hot dogs and 12 hamburgers were eaten at the picnic. What is the ratio of hot dogs to hamburgers?

   (c) The Colts won 40 football games and lost 60. What is the ratio of wins to losses?

(d) PROPORTION—(a) A cook needs 2 pounds of meat to feed 4 people and 8 pounds of meat to feed 16 people. What is the proportion for the number of pounds of meat to the number of people?

   (b) 3 waiters serve 9 tables at The Plaza Cafe. 6 waiters serve 18 tables in The Plaza Dining Room. Write the proportion for the number of waiters to the number of tables they serve.

PART—Each part is called $\frac{1}{8}$

SHARE—Each investor's share was $\frac{1}{4}$

RATIO—(a) The ratio of trucks to cars is $\frac{6}{11}$
A ratio compares two like quantities. Here you compare motor vehicles. The number you read first (trucks) is the numerator, the other number (cars) is the denominator.

(b) The ratio of hot dogs to hamburgers is $\frac{8}{12} = \frac{2}{3}$
Ratios can be reduced to lowest terms without changing the value. For every 2 hot dogs eaten, 3 hamburgers were eaten.

(c) The ratio of wins to losses is $\frac{40}{60} = \frac{4}{6} = \frac{2}{3}$
For every two games the Colts won, they lost 3 games.

PROPORTION—(a) The proportion is: number of pounds of meat to number of people

$$\text{pounds of meat} \quad \frac{2}{4} = \frac{8}{16} \quad \text{pounds of meat}$$
$$\text{people served} \qquad\qquad \text{people served}$$

A proportion shows that two ratios are equal.

(b) The proportion is:

$$\underset{\text{tables they serve}}{\overset{\text{waiters}}{\frac{3}{9}}} = \underset{\text{tables they serve}}{\overset{\text{waiters}}{\frac{6}{18}}}$$

The proportion is correct because $\frac{3}{9} = \frac{6}{18}$.

14.1. Word Problems

1. Jakes Used Car Lot was marked off into eight equal sections. What is the fraction name for each equal part?
    (1) 8   **(2)** $\frac{5}{8}$ **(3)** $\frac{1}{8}$ **(4)** $\frac{8}{10}$
    (5) none of these

    1. 1   2   3   4   5

2. An office has 11 push-button phones and 16 dial phones. What is the ratio of push-button to dial telephones?
    (1) 5   **(2)** $\frac{11}{16}$ **(3)** $1\frac{1}{16}$ **(4)** $\frac{16}{11}$
    (5) none of these

    2. 1   2   3   4   5

3. To make a simple salad dressing mix 1 tablespoon of vinegar with 3 tablespoons of oil. What is the ratio of vinegar to oil?
    (1) 2   **(2)** 4   **(3)** $\frac{3}{1}$ **(4)** $\frac{1}{3}$
    (5) none of these

    3. 1   2   3   4   5

4. 10 people drink 20 cups of tea at home. 4 people drink 8 cups of tea at the office. What is the proportion for the number of people to the number of cups of tea?
    (1) $\frac{10}{8} = \frac{4}{20}$ **(2)** $\frac{10}{20} = \frac{4}{8}$ **(3)** $\frac{2}{5} = \frac{2}{5}$ **(4)** $\frac{4}{8} = \frac{1}{2}$
    (5) none of these

    4. 1   2   3   4   5

5. 15 minutes is $\frac{1}{4}$ of an hour. What is the fraction name for 45 minutes? There are 60 minutes in 1 hour.
    (1) $\frac{1}{2}$ **(2)** $\frac{2}{4}$ **(3)** $\frac{2}{3}$ **(4)** $\frac{3}{4}$
    (5) none of these

    5. 1   2   3   4   5

15. ANSWERS

3.1. (a) $\frac{4}{6}$ (b) $\frac{4}{10}$

3.2. (a) $\frac{2}{4}$ (b) $\frac{2}{3}$ (c) $\frac{1}{2}$ (d) $\frac{3}{12}$ (e) $\frac{6}{8}$ (f) $\frac{2}{5}$

   (g) $\frac{1}{2}$ (b) $\frac{2}{6}$

3.5. **(a)** proper   **(b)** improper   **(c)** improper   **(d)** improper   **(e)** proper

4.1. **(a)** $\frac{3}{4}$ **(b)** $\frac{2}{6}$ **(c)** $\frac{5}{8}$ **(d)** $\frac{5}{8}$ **(e)** $\frac{2}{3}$ **(f)** $\frac{1}{2}$

     (g) $\frac{4}{5}$ **(h)** $\frac{3}{4}$

5.1. $\frac{1}{3} = \frac{2}{6}$

5.2. **(a)** $\frac{1}{2}$ **(b)** $\frac{2}{4}$ **(c)** $\frac{4}{8}$ **(d)** $\frac{3}{6}$ **(e)** $\frac{6}{12}$

6.1. **(a)** $\frac{2}{4}$ **(b)** $\frac{4}{8}$ **(c)** $\frac{2}{6}$ **(d)** $\frac{1}{2}$ **(e)** $\frac{6}{12}$

7.1. $\frac{6}{8} = \frac{3}{4}$

7.3. **(a)** $\frac{1}{3}$ **(b)** $\frac{1}{10}$ **(c)** $\frac{1}{2}$ **(d)** $\frac{1}{5}$ **(e)** $\frac{2}{3}$ **(f)** $\frac{1}{3}$

     (g) $\frac{1}{2}$ **(h)** $\frac{3}{4}$

8.1. $\frac{1}{3} = \frac{4}{12}$

9.1. **(a)** $5 = \frac{5}{1}$ **(b)** $\frac{6}{1}$ **(c)** $\frac{8}{1}$ **(d)** $\frac{12}{1}$ **(e)** $\frac{13}{1}$

11.1. $\frac{7}{3} = 2\frac{1}{3}$

11.4. **(a)** $2\frac{2}{3}$ **(b)** 5 **(c)** $3\frac{1}{2}$ **(d)** $2\frac{2}{7}$ **(e)** 2

12.1. $3\frac{1}{2} = \frac{7}{2}$

12.2. **(a)** $\frac{9}{2}$ **(b)** $\frac{13}{4}$ **(c)** $\frac{28}{5}$ **(d)** $\frac{55}{8}$ **(e)** $\frac{29}{10}$ **(f)** $\frac{11}{3}$

     (g) $\frac{24}{9}$ **(h)** $\frac{19}{5}$

14.1. Word Problems

     1. (3) $\frac{1}{8}$

     2. (2) $\frac{11}{16}$

     3. (4) $\frac{1}{3}$

     4. (2) $\frac{10}{20} = \frac{4}{8}$

     5. (4) $\frac{45}{60} = \frac{3}{4}$

TWELVE

Multiplication of Fractions

1. MULTIPLYING FRACTIONS

Multiplying fractions is fast and the steps are easy. The sign × tells you to multiply.

$\frac{3}{4} \times \frac{1}{2}$

Multiply numerators: $3 \times 1 = 3$.

Multiply denominators: $4 \times 2 = 8$.

Your answer is $\frac{3}{8}$.

$$\frac{3}{4} \times \frac{1}{2} = \frac{3}{}$$

$$\frac{3}{4} \times \frac{1}{2} = \frac{3}{8}$$

1.1. Test Yourself

Multiply: $\frac{2}{7} \times \frac{3}{5}$

Check your answer. If it is right, go to section 1.2. If it is wrong or you don't know what to do, go to section 1.1A.

1.1A. Multiply the numerators: the answer makes the numerator of the fraction you are looking for. This numerator is $2 \times 3 = 6$. Then multiply the denominators: the answer is the denominator of your fraction.

Finish your multiplication, check it, and go to section 1.2.

1.2. Sometimes your answer needs to be reduced to lowest terms.

Example: $\frac{2}{3} \times \frac{3}{4}$

Multiply the numerators: $2 \times 3 = 6$.

Multiply the denominators: $3 \times 4 = 12$.
Your answer is $\frac{6}{12}$.

Reduce to lowest terms.
Your final answer is $\frac{1}{2}$.

$$\frac{2}{3} \times \frac{3}{4} = \frac{6}{}$$

$$\frac{2}{3} \times \frac{3}{4} = \frac{6}{12}$$

$$\frac{6 \div 6}{12 \div 6} = \frac{1}{2}$$

If you don't remember how to reduce a fraction to lowest terms, go back to Chapter 11.

1.3. Practice

Multiply and reduce to lowest terms if necessary.

(a) $\frac{3}{4} \times \frac{1}{2}$ **(b)** $\frac{1}{2} \times \frac{1}{2}$ **(c)** $\frac{1}{2} \times \frac{1}{3}$ **(d)** $\frac{2}{3} \times \frac{1}{4}$ **(e)** $\frac{2}{3} \times \frac{3}{7}$

Check your answers.

2. MULTIPLYING MORE THAN TWO FRACTIONS

$\frac{1}{3} \times \frac{2}{5} \times \frac{2}{3}$

Multiply first two numerators: 1 × 2 = 2.

2
$\frac{1}{3} \times \frac{2}{5} \times \frac{2}{3} =$

Multiply answer and last numerator: 2 × 2 = 4.

4
2
$\frac{1}{3} \times \frac{2}{5} \times \frac{2}{3} = 4/$

Multiply first two denominators: 3 × 5 = 15.

$\frac{1}{3} \times \frac{2}{5} \times \frac{2}{3} = 4/$
15

Multiply answer and last denominator: 15 × 3 = 45.

$\frac{1}{3} \times \frac{2}{5} \times \frac{2}{3} = \frac{4}{45}$
15
45

Your answer is $\frac{4}{45}$.

2.1. Practice

Multiply and reduce answers to lowest terms, if necessary.

(a) $\frac{1}{3} \times \frac{2}{5} \times \frac{2}{7}$ **(b)** $\frac{3}{5} \times \frac{4}{7} \times \frac{1}{2}$ **(c)** $\frac{1}{2} \times \frac{1}{4} \times \frac{3}{4}$

(d) $\frac{5}{7} \times \frac{1}{2} \times \frac{1}{3}$ **(e)** $\frac{3}{4} \times \frac{1}{2} \times \frac{3}{7}$

Check your answers.

3. CANCELLING: A SHORT-CUT

Cancelling makes multiplying fractions easier. You cancel before you multiply.

Example: $\frac{3}{8} \times \frac{4}{5}$

Look at the numerators and denominators:
4 and 8 can both be divided by 4 with no remainder.

$\frac{3}{8} \times \frac{4}{5} =$

Divide 4 by 4: $4 \div 4 = 1$
Divide 8 by 4: $8 \div 4 = 2$
Multiply the numerators: $3 \times 1 = 3$.
Multiply the denominators: $2 \times 5 = 10$.
Your answer is $\frac{3}{10}$.

$$\frac{\overset{}{\underset{2}{3}}}{8} \times \frac{\overset{1}{4}}{5} =$$

$$\frac{3}{2} \times \frac{1}{5} = \frac{3}{10}$$

This way you don't need to reduce your answer to lowest terms: it is already in lowest terms.

Rule for cancelling: *each time you cancel a numerator, you must cancel a denominator.*

Example: $\frac{6}{7} \times \frac{5}{12}$

6 can be divided by 6 with no remainder.
12 can also be divided by 6 with no remainder.
If you cancel 6, you must also cancel 12.

$6 \div 6 = 1$

$12 \div 6 = 2$

$$\frac{6}{7} \times \frac{5}{12} =$$

$$\frac{\overset{1}{6}}{7} \times \frac{5}{\underset{2}{12}} =$$

$$\frac{1}{7} \times \frac{5}{2} = \frac{5}{14}$$

Multiply numerators and multiply denominators.
Your answer: $\frac{5}{14}$.

3.1. You may cancel more than once.
Example: $\frac{5}{6} \times \frac{9}{10}$

6 can be divided by 3 with no remainder.
9 can also be divided by 3 with no remainder.
Cancel:
$6 \div 3 = 2$
$9 \div 3 = 3$
5 can be divided by 5 with no remainder.
10 can also be divided by 5 with no remainder.
Cancel:
$5 \div 5 = 1$
$10 \div 5 = 2$

$$\frac{5}{6} \times \frac{9}{10} =$$

$$\frac{5}{\underset{2}{6}} \times \frac{\overset{3}{9}}{10} =$$

$$\frac{5}{2} \times \frac{3}{10} =$$

$$\frac{\overset{1}{5}}{2} \times \frac{3}{\underset{2}{10}} =$$

Multiply numerators and multiply denominators.
Your answer is $\frac{3}{4}$.

$$\frac{1}{2} \times \frac{3}{2} = \frac{3}{4}$$

3.2. Practice

Cancel, if possible, and multiply.

    (a) $\frac{3}{8} \times \frac{2}{5}$ **(b)** $\frac{2}{3} \times \frac{4}{7}$ **(c)** $\frac{4}{15} \times \frac{3}{5}$ **(d)** $\frac{4}{5} \times \frac{5}{7}$ **(e)** $\frac{2}{7} \times \frac{3}{4}$

Check your answers.

4. MULTIPLYING A FRACTION AND A WHOLE NUMBER

⅔ of a 6-foot board is made into a shelf. How long is the shelf?

⅔ × 6 feet

6 feet

Change 6 to a fraction:

6 = 6/1

| ⅔ × 6 = |
| ⅔ × 6/1 = |
| ⅔ × 6/1 = |

Cancel:

3 ÷ 3 = 1

6 ÷ 3 = 2

Multiply numerators and multiply denominators.

Your answer is 4/1.

It's an improper fraction. Change it to a whole number.

Your final answer: 4.

The shelf is 4 feet long.

| 2/1 × 2/1 = 4/1 |
| 4/1 = 4 |

4.1. Test Yourself

Multiply: 4 × 5/6 and change the answer to a whole number or a mixed number, if necessary.

Check your answer. If it is right, go to section 4.2. If it is wrong or you don't know what to do, go to section 4.1A.

4.1A. The first thing to do is to change 4 to a fraction. 4 = 4/1

Your multiplication now is: 4/1 × 5/6

Cancel if you can, then multiply numerators and denominators. Change your answer to a mixed number.

Check your answer.

4.2. Practice

Cancel, if possible, and multiply. If necessary, change your answers to a whole number or a mixed number, or reduce to lowest terms.

　　(a) ¾ × 5　　　　**(b)** ⅔ × 9　　　　**(c)** 4 × ⅖　　　　**(d)** ½ × 3　　　　**(e)** ⅝ × 2

Check your answers.

5. MULTIPLYING A FRACTION AND A MIXED NUMBER

There were $4\frac{1}{2}$ gallons of ice cream for the party but $\frac{2}{3}$ of it was left over. How much was left over?

$4\frac{1}{2} \times \frac{2}{3}$

Change $4\frac{1}{2}$ to an improper fraction:
$4 \times 2 = 8;\ 8 + 1 = 9$
$4\frac{1}{2} = \frac{9}{2}$
Cancel. Each time you cancel a numerator you must cancel a denominator.
$9 \div 3 = 3;\ 3 \div 3 = 1$
$2 \div 2 = 1;\ 2 \div 2 = 1$
Multiply numerators, multiply denominators.
Your answer: $\frac{3}{1}$.
Change it to a whole number:
Your final answer: 3.
There were 3 gallons of ice cream left over.

$4\frac{1}{2} \times \frac{2}{3} =$

$\frac{9}{2} \times \frac{2}{3} =$

$\frac{\overset{3}{\cancel{9}}}{\underset{1}{\cancel{2}}} \times \frac{\overset{1}{\cancel{2}}}{\underset{1}{\cancel{3}}} =$

$\frac{3}{1} \times \frac{1}{1} = \frac{3}{1}$

$\frac{3}{1} = 3$

5.1. Test Yourself

Multiply $5\frac{1}{2} \times \frac{2}{5}$
Check your answer. If it is right, go to section 6. If it is wrong or you don't know what to do, go to section 5.1A.
5.1A. The first thing to do is to change $5\frac{1}{2}$ to a fraction.

$5\frac{1}{2} = \frac{11}{2}$

If you don't remember how to change a mixed number to an improper fraction, go back to Chapter 11.
Now your multiplication is: $\frac{11}{2} \times \frac{2}{5}$
Cancel and multiply. Change your answer to a mixed number.

6. MULTIPLYING FRACTIONS, WHOLE NUMBERS, AND MIXED NUMBERS AT THE SAME TIME

The steps are always the same:
a. Change everything to fractions.
b. Cancel if you can.
c. Multiply numerators and multiply denominators.
d. Change your answer if it needs to be changed.

Example: $\frac{2}{5} \times 3\frac{1}{2} \times 10 \times \frac{6}{7}$

a. Change everything to fractions:
$$3\frac{1}{2} = \frac{7}{2}$$
$$10 = \frac{10}{1}$$

b. Cancel everything you can:
   2 and 2
   5 and 10
   7 and 7

c. Multiply numerators, multiply denominators:
$$1 \times 1 \times 2 \times 6 = 12$$
$$1 \times 1 \times 1 \times 1 = 1$$
   Your answer is $\frac{12}{1}$.

d. Change it to a whole number.
   Your final answer: 12.

$\frac{2}{5} \times 3\frac{1}{2} \times 10 \times \frac{6}{7} =$

$\frac{2}{5} \times \frac{7}{2} \times \frac{10}{1} \times \frac{6}{7} =$

$\frac{1}{\cancel{2}} \times \frac{\cancel{7}}{\cancel{2}} \times \frac{\cancel{10}^2}{1} \times \frac{\cancel{6}}{\cancel{7}} =$

$\frac{1}{1} \times \frac{1}{1} \times \frac{2}{1} \times \frac{6}{1} = \frac{12}{1}$

$\frac{12}{1} = 12$

6.1. Practice

(a) $\frac{7}{8} \times 3\frac{1}{3}$ **(b)** $3\frac{2}{5} \times 1\frac{1}{4}$ **(c)** $6 \times 3\frac{2}{3}$ **(d)** $5 \times 2\frac{1}{2} \times \frac{3}{4}$ **(e)** $7\frac{1}{3} \times 9$

Check your answers.

7. YOU CAN MULTIPLY ANYTHING

In this chapter, you have been learning to multiply any kind of fraction, mixed numbers, and whole numbers. All you have to do is remember the rules you have learned so far.

8. MULTIPLICATION REMINDERS

Before multiplying, change whole numbers to fractions,
                    change mixed numbers to fractions.
Before multiplying, cancel everything you can.
Each time you cancel a numerator, you must also cancel a denominator.
When multiplying, multiply numerators by numerators,
                    multiply denominators by denominators.
After multiplying, change your answer if it needs to be changed:
change improper fractions to whole numbers or mixed numbers;
reduce proper fractions to lowest terms.

8.1. More Practice

(a) $\frac{3}{13} \times \frac{1}{2}$ (b) $4\frac{5}{11} \times \frac{5}{7}$ (c) $1\frac{8}{32} \times 2\frac{7}{8}$ (d) $\frac{4}{7} \times \frac{3}{9}$

(e) $6 \times \frac{5}{12}$ (f) $3\frac{1}{4} \times 1\frac{2}{26}$ (g) $2\frac{2}{14} \times 1\frac{2}{5}$ (h) $\frac{6}{9} \times \frac{6}{9}$

(i) $\frac{8}{9} \times \frac{1}{2} \times \frac{3}{4}$ (j) $\frac{5}{12} \times 4 \times 1\frac{1}{15}$

Check your answers.

8.2. Word Problems

Before starting this section, review the multiplication word clues in Chapter 6. Here is another word clue for the multiplication of fractions:

OF—⅓ OF the 15-foot wall was finished. How much is that?

They ran ¾ OF the way around a 2½-mile track. How far did they run?

The word "of" in a word problem with fractions tells you to multiply.

$$\frac{1}{3} \text{ of } 15 = \frac{1}{3} \times 15 = \frac{1}{3} \times \frac{15}{1} = \frac{1}{\cancel{3}_{1}} \times \frac{\cancel{15}^{5}}{1} = \frac{5}{1} = 5$$

5 feet of the wall is finished.

$$\frac{3}{4} \text{ of } 2\frac{1}{2} = \frac{3}{4} \times 2\frac{1}{2} = \frac{3}{4} \times \frac{5}{2} = \frac{15}{8} = 1\frac{7}{8}.$$

They ran $1\frac{7}{8}$ miles.

1. Of the 56 office phones, ⅞ of them are push-button. How many telephones are push-button?
   (1) 42   (2) 49   (3) 45   (4) 54
   (5) none of these

2. One package of frozen strawberries weighs 10 ounces. How many ounces does ½ of a package weigh?
   (1) 5   (2) $\frac{2}{10}$ (3) 4   (4) 2
   (5) none of these

3. A T.V. special was 1½ hours long. Nick watched ⅔ of the special. How many hours did Nick watch T.V.?
   (1) ½ hour (2) 55 minutes (3) 1 hour (4) 45 minutes
   (5) none of these

4. Brian traveled 5¼ hours at a speed of 60 miles per hour. How many miles did he go?
   (1) 235   (2) 315   (3) 305   (4) 250
   (5) none of these

5. The highway from Franklin to Walton is $8\frac{1}{4}$ miles long. If $\frac{2}{3}$ of the road is under construction, how many miles of the road is under construction?
(1) $4\frac{1}{2}$ **(2)** $6\frac{1}{4}$ **(3)** 5 **(4)** $5\frac{1}{2}$
(5) none of these

6. If $\frac{3}{5}$ of 30 adults wear glasses, how many people in the group wear glasses?
(1) 20 **(2)** 15 **(3)** 16 **(4)** 18
(5) none of these

7. Mrs. Holmes planned to serve $\frac{1}{2}$ grapefruit to each of 6 people. How many whole grapefruits does she need?
(1) 3 **(2)** 5 **(3)** 4 **(4)** 2
(5) none of these

8. A lake is $\frac{1}{5}$ of a mile wide at its widest point. In the middle of the lake (half way across) there is a dock. What part of a mile is it from the shore to the dock?
(1) $\frac{1}{7}$ **(2)** $\frac{1}{10}$ **(3)** $\frac{1}{3}$ **(4)** $\frac{1}{2}$
(5) none of these

Check your answers. If all your answers are right, start Chapter 13. If one or two are wrong, make sure that you understand the right answers and then start Chapter 13. If you made more than two mistakes, you should read Chapter 12 again before starting Chapter 13.

9. ANSWERS

1.1. $\frac{6}{35}$

1.3. **(a)** $\frac{3}{8}$ **(b)** $\frac{1}{4}$ **(c)** $\frac{1}{6}$ **(d)** $\frac{2}{12} = \frac{1}{6}$ **(e)** $\frac{6}{21} = \frac{2}{7}$

2.1. **(a)** $\frac{4}{105}$ **(b)** $\frac{12}{70} = \frac{6}{35}$ **(c)** $\frac{3}{32}$ **(d)** $\frac{5}{42}$ **(e)** $\frac{9}{56}$

3.2. **(a)** $\frac{3}{20}$ **(b)** $\frac{8}{21}$ **(c)** $\frac{4}{25}$ **(d)** $\frac{4}{7}$ **(e)** $\frac{1}{2}$

4.1. $4 \times \frac{5}{6} = \frac{4}{1} \times \frac{5}{6} = \frac{10}{3} = 3\frac{1}{3}$

4.2. **(a)** $\frac{15}{4} = 3\frac{3}{4}$ **(b)** 6 **(c)** $1\frac{3}{5}$ **(d)** $1\frac{1}{2}$ **(e)** $1\frac{1}{4}$

5.1. $5\frac{1}{2} \times \frac{2}{5} = \frac{11}{2} \times \frac{2}{5} = \frac{11}{5} = 2\frac{1}{5}$

6.1. **(a)** $2\frac{11}{12}$ **(b)** $4\frac{1}{4}$ **(c)** 22 **(d)** $9\frac{3}{8}$ **(e)** 66

8.1. **(a)** $\frac{3}{26}$ **(b)** $3\frac{2}{11}$ **(c)** $2\frac{13}{16}$ **(d)** $\frac{4}{21}$ **(e)** $2\frac{1}{2}$ **(f)** $3\frac{1}{2}$
(g) 3 **(h)** $\frac{4}{9}$ **(i)** $\frac{1}{3}$ **(j)** $1\frac{7}{9}$

8.2. Word Problems

1. (2) $\frac{7}{8} \times 56 = \frac{7}{8} \times \frac{\overset{7}{56}}{1} = \frac{49}{1} = 49$

2. (1) $\frac{1}{2} \times 10 = \frac{1}{\underset{1}{2}} \times \frac{\overset{5}{10}}{1} = 5$

3. (3) $\frac{2}{3} \times 1\frac{1}{2} = \frac{\overset{1}{2}}{\underset{1}{3}} \times \frac{\overset{1}{3}}{\underset{1}{2}} = \frac{1}{1} = 1$

4. (2) $5\frac{1}{4} \times 60 = \frac{21}{\underset{1}{4}} \times \frac{\overset{15}{60}}{1} = \frac{315}{1} = 315$

5. (4) $\frac{2}{3} \times 8\frac{1}{4} = \frac{\overset{1}{2}}{\underset{1}{3}} \times \frac{\overset{11}{33}}{\underset{2}{4}} = \frac{11}{2} = 5\frac{1}{2}$

6. (4) $\frac{3}{5} \times 30 = \frac{3}{\underset{1}{5}} \times \frac{\overset{6}{30}}{1} = 18$

7. (1) $\frac{1}{2} \times 6 = \frac{1}{\underset{1}{2}} \times \frac{\overset{3}{6}}{1} = 3$

8. (2) $\frac{1}{2} \times \frac{1}{5} = \frac{1}{10}$

THIRTEEN

Division of Fractions

1. DIVIDING FRACTIONS

To divide fractions, what you really do is a multiplication.
Example: $\frac{1}{6} \div \frac{2}{7}$

Write your first fraction.

Change ÷ to ×

Turn your second fraction upside down.
Instead of $\frac{2}{7}$ you now have $\frac{7}{2}$

Instead of a division, you have to do a multiplication.
You already know how to do a multiplication.
Here there is nothing to cancel.
Multiply numerators and multiply denominators.
Your answer is $\frac{7}{12}$.

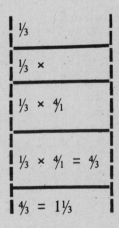

$\frac{1}{6}$

$\frac{1}{6}$ ×

$\frac{1}{6}$ × $\frac{7}{2}$ =

$\frac{1}{6}$ × $\frac{7}{2}$ = $\frac{7}{12}$

Here is another example: $\frac{1}{3} \div \frac{1}{4}$

Write the first fraction.

Change ÷ to ×.

Turn the second fraction upside down.
You now have a multiplication.

Multiply numerators and multiply denominators.
Your answer is $\frac{4}{3}$.
Change it to a mixed number.
Final answer: $1\frac{1}{3}$

$\frac{1}{3}$

$\frac{1}{3}$ ×

$\frac{1}{3}$ × $\frac{4}{1}$

$\frac{1}{3}$ × $\frac{4}{1}$ = $\frac{4}{3}$

$\frac{4}{3}$ = $1\frac{1}{3}$

1.1. Test Yourself

Do this division: $\frac{3}{4} \div \frac{2}{5}$

Check your answer. If it is right, go to section 1.2. If it is wrong or you don't know what to do, go to section 1.1A.

1.1A. The first thing to do is to change your division into a multiplication. For that, keep the first fraction, change ÷ to ×, and turn the second fraction upside down. Your multiplication is:

$$\frac{3}{4} \times \frac{5}{2}$$

You know how to do a multiplication.
Check your answer.

1.2. Practice

    (a) $\frac{2}{3} \div \frac{1}{2}$ **(b)** $\frac{4}{5} \div \frac{1}{3}$ **(c)** $\frac{6}{7} \div \frac{5}{6}$

    (d) $\frac{7}{8} \div \frac{4}{5}$ **(e)** $\frac{4}{9} \div \frac{1}{4}$

Check your answers.

2. CANCELLING

After you have changed your division to a multiplication, you may be able to cancel.
Example: $\frac{1}{2} \div \frac{1}{4}$
Write the first fraction; change ÷ to ×.

Turn the second fraction upside down.
You now have your multiplication.
Cancel everything you can.

Multiply the fractions.
Your answer: $\frac{2}{1}$
Change it to a whole number.
Your final answer: 2

$\frac{1}{2} \times$
$\frac{1}{2} \times \frac{4}{1} =$
$\frac{1}{2} \times \overset{2}{\underset{1}{4}}{}_{1} =$
$\frac{1}{1} \times \frac{2}{1} = \frac{2}{1}$
$\frac{2}{1} = 2$

2.1. Test Yourself

Do this division: $\frac{2}{3} \div \frac{4}{9}$
Check your answer. If it is right, go to section 2.2. If it is wrong or you don't know what to do, go to section 2.1A.

2.1A. First, change your division to a multiplication.

Write your first fraction, change ÷ to ×, and turn your second fraction upside down.

Your multiplication is: $\frac{2}{3} \times \frac{9}{4}$

Before multiplying, cancel everything you can.

Then multiply your fractions.

Change your answer to a mixed number.

Check your answer.

2.2. Practice

(a) $\frac{2}{5} \div \frac{2}{3}$ **(b)** $\frac{3}{7} \div \frac{3}{4}$ **(c)** $\frac{1}{2} \div \frac{1}{8}$

(d) $\frac{4}{9} \div \frac{2}{3}$ **(e)** $\frac{5}{8} \div \frac{3}{4}$

Check your answers.

3. DIVIDING A FRACTION BY A WHOLE NUMBER, OR A WHOLE NUMBER BY A FRACTION

Change the whole number to a fraction. All the other steps are the same.

Example: $\frac{3}{5} \div 2$

Change the whole number to a fraction

$2 = \frac{2}{1}$

Change ÷ to × and turn the second fraction upside down.

Nothing can be cancelled.

Multiply the fractions.

Your answer is $\frac{3}{10}$.

Example: $6 \div \frac{3}{4}$

Change the whole number to a fraction

$6 = \frac{6}{1}$

Change ÷ to × and turn the second fraction upside down.

Cancel everything you can (each time you cancel a numerator, you must cancel a denominator).

Multiply the fractions. Your answer is $\frac{8}{1}$.

Change it to a whole number.

Final answer: 8.

$\frac{3}{5} \div \frac{2}{1} =$

$\frac{3}{5} \times \frac{1}{2} =$

$\frac{3}{5} \times \frac{1}{2} = \frac{3}{10}$

$\frac{6}{1} \div \frac{3}{4} =$

$\frac{6}{1} \times \frac{4}{3} =$

$\overset{2}{\frac{6}{1}} \times \frac{4}{\underset{1}{3}} =$

$\frac{2}{1} \times \frac{4}{1} = \frac{8}{1}$

$\frac{8}{1} = 8$

3.1. Practice

(a) $\frac{3}{8} \div 3$ **(b)** $\frac{4}{7} \div 2$ **(c)** $\frac{1}{3} \div 6$ **(d)** $\frac{5}{6} \div 5$ **(e)** $\frac{7}{8} \div 8$

Check your answers.

4. DIVIDING A MIXED NUMBER BY A FRACTION, OR A FRACTION BY A MIXED NUMBER

Change the mixed number to an improper fraction.
All the other steps are the same.
Example: $2\frac{1}{2} \div \frac{3}{4}$
Change the mixed number to a fraction.

$$\frac{5}{2} \div \frac{3}{4} =$$

Change $\div$ to $\times$ and turn the second fraction upside down.

$$\frac{5}{2} \times \frac{4}{3} =$$

Cancel everything you can.

$$\frac{5}{\underset{1}{2}} \times \overset{2}{\frac{4}{3}} =$$

Multiply the fractions.
Your answer is $\frac{10}{3}$.

$$\frac{5}{1} \times \frac{2}{3} = \frac{10}{3}$$

Change it to a mixed number.
Final answer: $3\frac{1}{3}$

$$\frac{10}{3} = 3\frac{1}{3}$$

Example: $\frac{2}{5} \div 3\frac{1}{5}$
Change the mixed number to a fraction.

$$\frac{2}{5} \div \frac{16}{5} =$$

Change $\div$ to $\times$ and turn second fraction upside down.

$$\frac{2}{5} \times \frac{5}{16} =$$

Cancel everything you can (each time you cancel a numerator, you must cancel a denominator).

$$\overset{1}{\underset{1}{\frac{2}{5}}} \times \overset{1}{\underset{8}{\frac{5}{16}}} =$$

Multiply the fractions.
Your answer is $\frac{1}{8}$.

$$\frac{1}{1} \times \frac{1}{8} = \frac{1}{8}$$

4.1. Practice

(a) $4\frac{1}{2} \div \frac{3}{4}$ **(b)** $1\frac{1}{2} \div \frac{1}{8}$ **(c)** $\frac{2}{3} \div 2\frac{1}{9}$

(d) $3\frac{3}{4} \div \frac{5}{7}$ **(e)** $1\frac{5}{8} \div \frac{7}{9}$

Check your answers.

5. DIVIDING A MIXED NUMBER BY A MIXED NUMBER

Change the mixed numbers to improper fractions.
All the other steps are the same.
Example: $3\frac{1}{4} \div 2\frac{3}{4}$
Change the mixed numbers to improper fractions.
Change $\div$ to $\times$ and turn the second fraction upside down.

Cancel everything you can.

Multiply the fractions.
Your answer: $\frac{13}{11}$.
Change it to a mixed number.
Final answer: $1\frac{2}{11}$

$$\frac{13}{4} \div \frac{11}{4} =$$

$$\frac{13}{4} \times \frac{4}{11} =$$

$$\frac{13}{4} \times \overset{1}{\frac{4}{11}} =$$

$$\frac{13}{1} \times \frac{1}{11} = \frac{13}{11}$$

$$\frac{13}{11} = 1\frac{2}{11}$$

5.1. Practice

(a) $5\frac{1}{3} \div 1\frac{1}{3}$ (b) $4\frac{1}{2} \div 1\frac{1}{2}$ (c) $3\frac{3}{5} \div 2\frac{2}{5}$

(d) $4\frac{4}{9} \div 1\frac{1}{9}$ (e) $2\frac{5}{8} \div 3\frac{3}{8}$

Check your answers.

6. DIVIDING A MIXED NUMBER BY A WHOLE NUMBER, OR A WHOLE NUMBER BY A MIXED NUMBER

Change the whole number and the mixed number to improper fractions. All the other steps are the same.
Example: $5\frac{1}{3} \div 4$
Change the mixed number and the whole number to improper fractions.

Change $\div$ to $\times$ and turn second fraction upside down.

Cancel everything you can.

Multiply the fractions. Your answer: $\frac{4}{3}$

Change it to a mixed number.
Your final answer: $1\frac{1}{3}$.

$$\frac{16}{3} \div \frac{4}{1} =$$

$$\frac{16}{3} \times \frac{1}{4} =$$

$$\overset{4}{\frac{16}{3}} \times \frac{1}{\underset{1}{4}} =$$

$$\frac{4}{3} \times \frac{1}{1} = \frac{4}{3}$$

$$\frac{4}{3} = 1\frac{1}{3}$$

Example: $3 \div 2\frac{1}{2}$

Change the whole number and the mixed number to improper fractions.

$\frac{3}{1} \div \frac{5}{2} =$

Change $\div$ to $\times$ and turn second fraction upside down.

$\frac{3}{1} \times \frac{2}{5} =$

There is nothing to cancel.
Multiply the fractions. Your answer: $\frac{6}{5}$

$\frac{3}{1} \times \frac{2}{5} = \frac{6}{5}$

Change it to a mixed number. Your final answer: $1\frac{1}{5}$.

$\frac{6}{5} = 1\frac{1}{5}$

6.1. Practice

(a) $5 \div 2\frac{1}{2}$ (b) $7\frac{1}{2} \div 5$ (c) $6 \div 4\frac{2}{3}$

(d) $10 \div 2\frac{3}{4}$ (e) $5\frac{2}{3} \div 3$

Check your answers.

7. YOU CAN DIVIDE ANYTHING

In this chapter, you have been learning how to divide any kind of fractions, mixed numbers, and whole numbers. Just remember the rules you have learned so far.

8. DIVISION REMINDERS

Before dividing, change whole numbers to fractions;
                    change mixed numbers to fractions.

Then keep the first fraction, change $\div$ to $\times$, and turn the second fraction upside down.

Before multiplying, cancel everything you can.

Each time you cancel a numerator, you must also cancel a denominator.

Multiply the fractions.

After multiplying, change your answer if it needs to be changed:

    change improper fractions to whole numbers or mixed numbers;
    reduce proper fractions to lowest terms.

8.1. More Practice

(a) $\frac{7}{18} \div \frac{21}{36} =$

(b) $\frac{3}{8} \div \frac{5}{6} =$

(c) $\frac{9}{10} \div \frac{54}{60} =$

(d) $5 \div \frac{5}{13} =$

(e) $12 \div \frac{4}{9} =$

(f) $16 \div \frac{24}{25} =$

(g) $3 \div 2\frac{1}{10} =$

(h) $6 \div 9\frac{3}{5} =$

(i) $7 \div 4\frac{3}{8} =$

(j) $\frac{4}{5} \div 28 =$

Check your answers.

8.2. Word Problems

Before starting this section, review the division word clues in Chapter 4.

1. How many ½-hour television shows do you see if you watch T.V. for 3 hours?
(1) 4　　　(2) 5　　　(3) 6　　　(4) 3
(5) none of these

1. 1 2 3 4 5

2. 45 pounds of potting soil were put into bags weighing 7½ pounds each. How many bags of soil were there?
(1) 6　　　(2) 7　　　(3) 8　　　(4) 5
(5) none of these

2. 1 2 3 4 5

3. A tank holds $\frac{1}{12}$ ton of oil. How many tanks are needed for ¾ tons of oil?
(1) 12　　　(2) 7　　　(3) 10　　　(4) 9
(5) none of these

3. 1 2 3 4 5

4. A road 6¾ miles long has a traffic sign every ⅜ of a mile. How many signs are there along the road?
(1) 15　　　(2) 18　　　(3) 17　　　(4) 16
(5) none of these

4. 1 2 3 4 5

5. 5⅓ acres of land are divided up among 8 people. What is each person's share?
(1) ⅔　　　(2) ⅓　　　(3) 1⅓　　　(4) 1
(5) none of these

5. 1 2 3 4 5

6. How many ⅑'s are there in ⅔'s?
(1) 4　　　(2) 7　　　(3) 3　　　(4) 9
(5) none of these

6. 1 2 3 4 5

7. A bus leaves Wallerville every 1½ hours. How many buses leave in 24 hours?
(1) 12　　　(2) 18　　　(3) 16　　　(4) 20
(5) none of these

7. 1 2 3 4 5

8. Yuki decorated a 12½-foot wall by painting a stripe every 2½ feet. How many stripes did Yuki paint on the wall?

    (1) 5   **(2)** 6   **(3)** 4   **(4)** 7

    (5) none of these

8.   1   2   3   4   5

Check your answers. If all your answers are right, start Chapter 14. If one or two are wrong, make sure you understand the right answers and then start Chapter 14. If you made more than two mistakes, you should read Chapter 13 again before you start Chapter 14.

9. ANSWERS

1.1. $\frac{3}{4} \div \frac{2}{5} = \frac{3}{4} \times \frac{5}{2} = \frac{15}{8} = 1\frac{7}{8}$

1.2. (a) $1\frac{1}{3}$ **(b)** $2\frac{2}{5}$ **(c)** $1\frac{1}{35}$ **(d)** $1\frac{3}{32}$ **(e)** $1\frac{7}{9}$

2.1. $\frac{2}{3} \div \frac{4}{9} = \frac{2}{3} \times \frac{9}{4} = \frac{3}{2} = 1\frac{1}{2}$

2.2. (a) $\frac{3}{5}$ **(b)** $\frac{4}{7}$ **(c)** 4   **(d)** $\frac{2}{3}$ **(e)** $\frac{5}{6}$

3.1. (a) $\frac{1}{8}$ **(b)** $\frac{2}{7}$ **(c)** $\frac{1}{18}$ **(d)** $\frac{1}{6}$ **(e)** $\frac{7}{64}$

4.1. (a) 6   **(b)** 12   **(c)** $\frac{6}{19}$ **(d)** $5\frac{1}{4}$ **(e)** $2\frac{5}{56}$

5.1. (a) $3\frac{1}{5}$ **(b)** 3   **(c)** $1\frac{1}{2}$ **(d)** 4   **(e)** $\frac{7}{9}$

6.1. (a) 2   **(b)** $1\frac{1}{2}$ **(c)** $1\frac{2}{7}$ **(d)** 4   **(e)** $1\frac{8}{9}$

8.1. (a) $\frac{2}{3}$ **(b)** $\frac{9}{20}$ **(c)** 1   **(d)** 13   **(e)** 27

    (f) $16\frac{2}{3}$ **(g)** $1\frac{3}{7}$ **(h)** $\frac{5}{8}$ **(i)** $1\frac{3}{5}$ **(j)** $\frac{1}{35}$

8.2. Word Problems

1. (3) $3 \div \frac{1}{2} = \frac{3}{1} \times \frac{2}{1} = \frac{6}{1} = 6$

2. (1) $45 \div 7\frac{1}{2} = \frac{\overset{3}{45}}{1} \times \frac{2}{\underset{1}{15}} = \frac{6}{1} = 6$

3. (4) $\frac{3}{4} \div \frac{1}{12} = \frac{3}{\underset{1}{4}} \times \frac{\overset{3}{12}}{1} = \frac{9}{1} = 9$

4. (2) $6\frac{3}{4} \div \frac{3}{8} = \frac{\overset{9}{27}}{\underset{1}{4}} \times \frac{\overset{2}{8}}{\underset{1}{3}} = \frac{18}{1} = 18$

5. (1) $5\frac{1}{3} \div 8 = \frac{\overset{2}{16}}{3} \times \frac{1}{\underset{1}{8}} = \frac{2}{3}$

6. (5) $\frac{2}{3} \div \frac{1}{9} = \frac{2}{\underset{1}{3}} \times \frac{\overset{3}{9}}{1} = \frac{6}{1} = 6$

7. (3) $24 \div 1\frac{1}{2} = \frac{\overset{8}{24}}{1} \times \frac{2}{\underset{1}{3}} = \frac{16}{1} = 16$

8. (1) $12\frac{1}{2} \div 2\frac{1}{2} = \frac{\overset{5}{25}}{\underset{1}{2}} \times \frac{\overset{1}{2}}{\underset{1}{5}} = \frac{5}{1} = 5$

FOURTEEN

Addition of Like Fractions

1. LIKE FRACTIONS

Fractions that have the same denominator are called LIKE FRACTIONS.

For example, $\frac{2}{3}$ and $\frac{1}{3}$ are like fractions.

$\frac{3}{4}$ and $\frac{7}{4}$ are like fractions.

2. UNLIKE FRACTIONS

Fractions that do not have the same denominator are called UNLIKE FRACTIONS.

For example, $\frac{3}{4}$ and $\frac{5}{6}$ are unlike fractions.

$\frac{3}{2}$ and $\frac{2}{5}$ are unlike fractions.

2.1 Practice

Say if the following fractions are like or unlike

(a) $\frac{7}{8}$ and $\frac{8}{7}$ **(b)** $\frac{6}{9}$ and $\frac{4}{9}$ **(c)** $\frac{11}{12}$ and $\frac{3}{12}$

(d) $\frac{1}{3}$ and $\frac{3}{5}$ **(e)** $\frac{2}{6}$ and $\frac{5}{6}$

Check your answers.

In this chapter and in the next chapter, we are going to work only with like fractions.

3. ADDING LIKE FRACTIONS

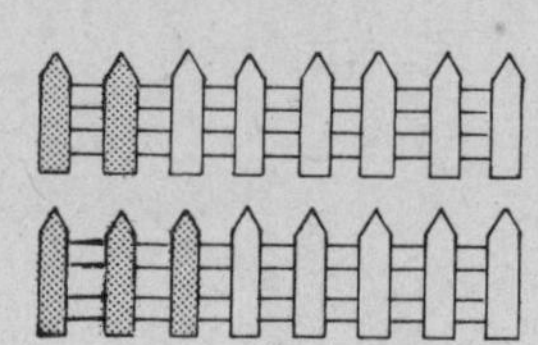

$\frac{2}{8}$ of a fence was painted in the morning, and $\frac{3}{8}$ of the fence was painted in the afternoon. What part of the fence was painted in the whole day?

$\frac{2}{8} + \frac{3}{8}$

To.add like fractions, add the numerators:

$2 + 3 = 5.$

Then just copy the denominator.

Your answer: $\frac{5}{8}$

$\frac{5}{8}$ of the fence was painted.

You can also use the column form:

$\frac{2}{8} + \frac{3}{8} = \frac{5}{}$

$\frac{2}{8} + \frac{3}{8} = \frac{5}{8}$

$\begin{array}{r} \frac{2}{8} \\ + \frac{3}{8} \\ \hline \frac{5}{} \end{array} \qquad \begin{array}{r} \frac{2}{8} \\ + \frac{3}{8} \\ \hline \frac{5}{8} \end{array}$

3.1. Practice

(a) $\frac{3}{7} + \frac{3}{7}$ **(b)** $\frac{2}{11} + \frac{4}{11}$ **(c)** $\frac{7}{15} + \frac{4}{15}$

(d) $\frac{3}{13} + \frac{8}{13}$ **(e)** $\frac{1}{5} + \frac{3}{5}$

Check your answers.

4. REDUCING THE SUM TO LOWEST TERMS

Sometimes the sum of two fractions can be reduced to lowest terms.

How people get to
work in a major U.S. City

What fraction of the population
takes city transportation to work?

$\frac{1}{4} + \frac{1}{4}$

Add the numerators.
Bring down the denominator.
Your answer is $\frac{2}{4}$.

$\begin{array}{r} \frac{1}{4} \\ + \frac{1}{4} \\ \hline \frac{2}{4} \end{array}$

$\frac{2}{4} = \frac{1}{2}$

Reduce it to lowest terms.
Your final answer is $\frac{1}{2}$.

When you reduce your answer to lowest terms, your final answer has a denominator
that is different from the denominator of the two fractions you added. Always give your
answers to fraction problems in lowest terms.

4.1. Practice

(a) $\frac{1}{6} + \frac{1}{6}$ **(b)** $\frac{5}{8} + \frac{1}{8}$ **(c)** $\frac{8}{15} + \frac{2}{15}$

(d) $\frac{4}{9} + \frac{2}{9}$ **(e)** $\frac{3}{14} + \frac{5}{14}$ Check your answers.

5. SUMS THAT ARE IMPROPER FRACTIONS

When the sum is an improper fraction, be sure to change it to a mixed number or a whole number.

The Browns and the Rodrigos went to the restaurant together. The Browns ate $\frac{6}{8}$ of a pizza, and the Rodrigos also ate $\frac{6}{8}$ of a pizza. What was the total amount of pizza eaten by the two families?

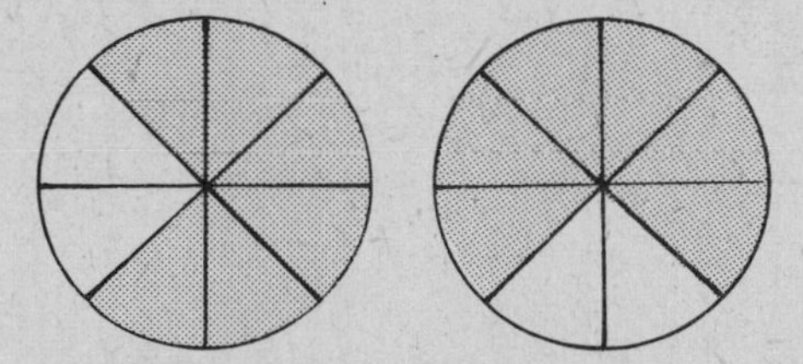

Brown Rodrigos

$\frac{6}{8} + \frac{6}{8}$

Add the numerators and copy the denominator.
Your answer is $\frac{12}{8}$.
It is an improper fraction.

$\frac{6}{8} + \frac{6}{8} = \frac{12}{8}$

Change it to a mixed number.
Your final answer: $1\frac{1}{2}$ pizzas.

$\frac{12}{8} = 1\frac{4}{8} = 1\frac{1}{2}$

Here is another example:
$\frac{3}{4}$ of the wall is already papered. $\frac{1}{4}$ is being papered now. What fraction of the wall is about to be finished?

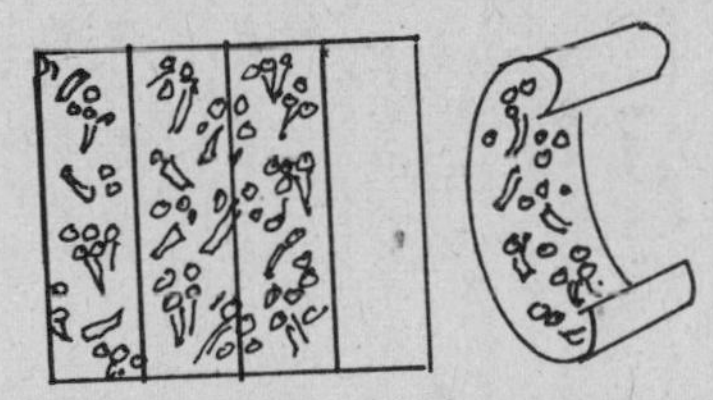

$\frac{3}{4} + \frac{1}{4}$

Add the numerators and copy the denominator.
Your answer is $\frac{4}{4}$.
It is an improper fraction.
Change it to a whole number.
Your final answer is 1. The whole wall is about to be finished.

$\frac{3}{4} + \frac{1}{4} = \frac{4}{4}$

$\frac{4}{4} = 1$

5.1. Practice

Add. Change the sum to a whole number or to a mixed number.

(a) $\frac{5}{6} + \frac{4}{6}$ (b) $\frac{7}{8} + \frac{3}{8}$ (c) $\frac{2}{3} + \frac{2}{3}$

(d) $\frac{4}{5} + \frac{3}{5}$ (e) $\frac{8}{16} + \frac{9}{16}$

Check your answers.

6. ADDING MORE THAN TWO LIKE FRACTIONS

You already know how to add three numbers or more. All you have to do to add three like fractions or more is add their numerators.
Add $\frac{3}{7} + \frac{6}{7} + \frac{4}{7}$
Add the first two numerators.

$$\begin{array}{r} \frac{3}{7} \\ \frac{6}{7} \quad 9 \\ + \frac{4}{7} \\ \hline \end{array}$$

Add the answer and the last numerator.
Bring down the denominator.

$$\begin{array}{r} \frac{3}{7} \\ \frac{6}{7} \quad 9 \\ + \frac{4}{7} \quad 13 \\ \hline \frac{13}{7} \end{array}$$

Your answer is $\frac{13}{7}$.
It is an improper fraction.
Change it to a mixed number.
Your final answer: $1\frac{6}{7}$.

$$\frac{13}{7} = 1\frac{6}{7}$$

6.1. Practice

Add. Change the sum if necessary. You may have to reduce it, or change it to a mixed number or to a whole number.

(a) $\frac{2}{9} + \frac{3}{9} + \frac{3}{9}$ (b) $\frac{3}{10} + \frac{5}{10} + \frac{1}{10}$

(c) $\frac{4}{12} + \frac{5}{12} + \frac{6}{12}$ (d) $\frac{9}{16} + \frac{4}{16} + \frac{2}{16}$

(e) $\frac{3}{18} + \frac{2}{18} + \frac{11}{18}$

Check your answers. If they are not all right, read section 6 again. Then go to section 7.

7. ADDING MIXED NUMBERS

A recipe calls for $1\frac{2}{8}$ cups of whole wheat flour and $1\frac{4}{8}$ cups of white flour. How much flour is that altogether?

$1\frac{2}{8} + 1\frac{4}{8}$ cups.

To add mixed numbers, put them in column form.
Put fractions under fractions, and units under units.

$$\begin{array}{r} 1\frac{2}{8} \\ + 1\frac{4}{8} \\ \hline \end{array}$$

Add the fractions first (add numerators; bring down the denominator).

$$\begin{array}{r} 1\frac{2}{8} \\ + 1\frac{4}{8} \\ \hline \frac{6}{8} \end{array}$$

Then add units.

Your answer: $2\frac{6}{8}$
Reduce the fraction to lowest terms.
Your final answer: $2\frac{3}{4}$
The recipe needs $2\frac{3}{4}$ cups of flour.

$$\begin{array}{r} 1\frac{2}{8} \\ + 1\frac{4}{8} \\ \hline 2\frac{6}{8} = 2\frac{3}{4} \end{array}$$

8. ADDING MORE THAN TWO MIXED NUMBERS

$3\frac{2}{9} + 2\frac{4}{9} + 6\frac{1}{9}$

Put in column form.
Add the fractions first (add the first and second numerators; then add the result and the third numerator; bring down the denominator).

$$\begin{array}{rcc} 3\frac{2}{9} & & \\ 2\frac{4}{9} & 6 & \\ + 6\frac{1}{9} & & 7 \\ \hline \frac{7}{9} & & \end{array}$$

Then add the units (add the first two numbers; add the result and the third number).
Your answer: $11\frac{7}{9}$
It doesn't need to be changed.

$$\begin{array}{rcc} & & 3\frac{2}{9} \\ & 5 & 2\frac{4}{9} \\ 11 & & + 6\frac{1}{9} \\ \hline & & 11\frac{7}{9} \end{array}$$

8.1. Practice

Add these mixed numbers. Reduce fractions when necessary.

(a) $3\frac{3}{8} + 6\frac{2}{8}$ (b) $9\frac{5}{12} + 6\frac{4}{12}$ (c) $2\frac{3}{15} + 5\frac{4}{15} + 4\frac{6}{15}$

(d) $1\frac{1}{9} + 5\frac{2}{9} + 2\frac{3}{9}$ (b) $12\frac{2}{14} + 5\frac{6}{14} + 2\frac{3}{14}$

Check your answers.

9. CARRYING OVER

You may have to carry over from the fractions column to the units column.

Example: $2\frac{5}{9} + 3\frac{8}{9}$

Put in column form.

Add the numerators and bring down the denominator.

$$\begin{array}{r} 2\frac{5}{9} \\ + 3\frac{8}{9} \\ \hline \frac{13}{9} \end{array}$$

Add the units.

Your answer is $5\frac{13}{9}$.

$\frac{13}{9}$ is an improper fraction.

Change it to a mixed number: $1\frac{4}{9}$.

$$\begin{array}{r} 2\frac{5}{9} \\ + 3\frac{8}{9} \\ \hline 5\frac{13}{9} \end{array}$$

Carry the 1 over to the units column.

Your final answer: $6\frac{4}{9}$.

$$\begin{array}{r} 1 \\ 2\frac{5}{9} \\ + 3\frac{8}{9} \\ \hline 5\frac{13}{9} \\ 6\frac{4}{9} \end{array}$$

9.1. Test Yourself

Add $5\frac{3}{8} + 10\frac{7}{8} =$

Check your answer. If it is right, go to section 9.2. If it is wrong, or you don't know what to do, go to section 9.1A.

9.1A. Put your numbers in column form and add them. Your final answer should be $15\frac{10}{8}$. Then you have to change $\frac{10}{8}$ to a mixed number, carry over, and reduce the remaining fraction to lowest terms. Finish and check your answer. Then go to section 9.2.

9.2. Practice

(a) $\begin{array}{r} 5\frac{4}{5} \\ + 2\frac{3}{5} \\ \hline \end{array}$

(b) $\begin{array}{r} 7\frac{8}{13} \\ + 4\frac{10}{13} \\ \hline \end{array}$

(c) $\begin{array}{r} 15\frac{2}{4} \\ + 4\frac{3}{4} \\ \hline \end{array}$

(d) $6\frac{3}{6} + 8\frac{4}{6} =$

(e) $7\frac{7}{12} + 10\frac{5}{12} =$

Check your answers.

10. ADDING MIXED NUMBERS, WHOLE NUMBERS, AND FRACTIONS

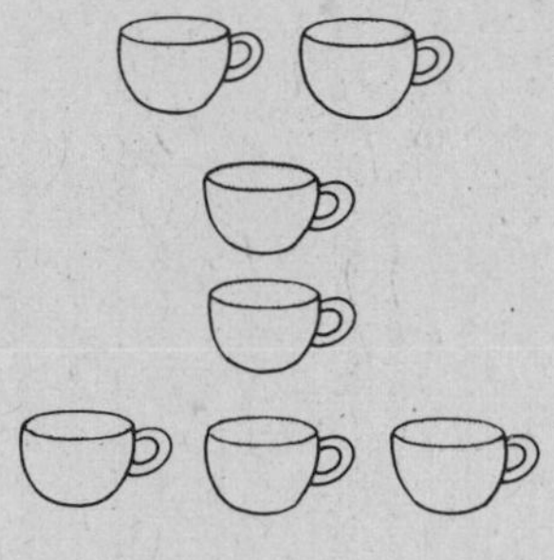

morning

break

lunch

evening

You drink $1\frac{1}{2}$ cups of coffee in the morning, $\frac{1}{2}$ cup at break, 1 cup at lunch, and $2\frac{1}{2}$ more cups of coffee in the evening. How much coffee do you drink all day?

$1\frac{1}{2} + \frac{1}{2} + 1 + 2\frac{1}{2}$ cups

Put in column form. Put fractions under fractions, units under units.

Add the numerators. Bring down the denominator.

Add the units.

Your answer: $4\frac{3}{2}$

$\frac{3}{2}$ is improper: change it to a mixed number.

$\frac{3}{2} = 1\frac{1}{2}$

Carry the 1 over to the units column.

Your final answer: $5\frac{1}{2}$.

You drink $5\frac{1}{2}$ cups of coffee.

units	fractions	
1	$\frac{1}{2}$	2
	$\frac{1}{2}$	
1		3
2	$\frac{1}{2}$	
4	$\frac{3}{2}$	

units	fractions
1	
1	$\frac{1}{2}$
	$\frac{1}{2}$
1	
2	$\frac{1}{2}$
4	$\frac{3}{2}$
5	$\frac{1}{2}$

10.1 Practice

(a) $2\frac{2}{5} + 3\frac{4}{5}$ **(b)** $9\frac{3}{8} + \frac{4}{8} + 6\frac{2}{8}$ **(c)** $3\frac{5}{6} + 4\frac{2}{6}$

(d) $4\frac{3}{7} + 6\frac{4}{7}$ **(e)** $2\frac{3}{4} + 1\frac{2}{4} + 4\frac{2}{4}$

Check your answers.

11. ADDITION REMINDERS

To add fractions, mixed numbers, and whole numbers, use the column form. Put fractions under fractions, units under units, tens under tens, and so on.

To add $1\frac{3}{9} + 12\frac{1}{9} + 3\frac{2}{9}$ write:

When adding fractions, add the numerators only. Denominators must all be the same.

Change improper fractions to mixed numbers

or whole numbers.

Carry over when necessary.

Reduce answers to lowest terms.

tens	units	fractions
	1	$\frac{3}{9}$
1	2	$\frac{1}{9}$
+	3	$\frac{2}{9}$

11.1. Word Problems

Before starting this section, review the addition word clues in Chapter 2.

1. An electrician needs the following amounts of wire to finish a job: $2\frac{1}{3}$ feet, $1\frac{2}{3}$ feet, $3\frac{2}{3}$ feet. How much wire does the electrician need in all?
   (1) $6\frac{2}{3}$ (2) $7\frac{1}{3}$ (3) $7\frac{2}{3}$ (4) $6\frac{1}{3}$
   (5) none of these

   1. 1 2 3 4 5

2. Your car burns oil. You put in $1\frac{3}{4}$ quarts on Monday, $2\frac{1}{4}$ quarts on Friday and $1\frac{1}{4}$ quarts the next Tuesday. How many quarts of oil did your car use?
   (1) 5   (2) $5\frac{1}{4}$ (3) $4\frac{3}{4}$ (4) $4\frac{6}{4}$
   (5) none of these

   2. 1 2 3 4 5

3. It takes $1\frac{5}{8}$ yards to sew a skirt, $\frac{5}{8}$ yard for a vest, and $2\frac{7}{8}$ yards for a jacket. How many yards of fabric do you need?
   (1) $4\frac{2}{8}$ (2) $4\frac{7}{8}$ (3) 5   (4) $5\frac{1}{8}$
   (5) none of these

   3. 1 2 3 4 5

4. A jogger ran $3\frac{9}{10}$ miles in the morning and another $\frac{6}{10}$ mile at night. How many miles all day?
   (1) $3\frac{1}{2}$ (2) $3\frac{1}{10}$ (3) $1\frac{1}{2}$ (4) $3\frac{4}{10}$
   (5) none of these

   4. 1 2 3 4 5

5. A cake recipe takes $\frac{3}{8}$ cup of nuts in the cake and $\frac{3}{8}$ cup of nuts on the frosting. How many cups of nuts in all?
   (1) $1\frac{2}{3}$ (2) $1\frac{1}{4}$ (3) $\frac{3}{4}$ (4) $\frac{3}{8}$
   (5) none of these

   5. 1 2 3 4 5

Check your answers. If all your answers are right, begin Chapter 15. If one or two are wrong, make sure that you understand the right answers before starting Chapter 15. If you made more than two mistakes, you should read Chapter 14 again before starting Chapter 15.

12. ANSWERS

2.1. **(a)** unlike   **(b)** like   **(c)** like   **(d)** unlike   **(e)** like

3.1. **(a)** $\frac{6}{7}$ **(b)** $\frac{9}{11}$ **(c)** $\frac{11}{15}$ **(d)** $\frac{11}{13}$ **(e)** $\frac{4}{5}$

4.1. **(a)** $\frac{1}{3}$ **(b)** $\frac{3}{4}$ **(c)** $\frac{2}{3}$ **(d)** $\frac{2}{3}$ **(e)** $\frac{4}{7}$

5.1. **(a)** $1\frac{3}{6} = 1\frac{1}{2}$ **(b)** $1\frac{1}{4}$ **(c)** $1\frac{1}{3}$ **(d)** $1\frac{2}{5}$ **(e)** $1\frac{1}{16}$

6.1. **(a)** $\frac{8}{9}$ **(b)** $\frac{9}{10}$ **(c)** $1\frac{1}{4}$ **(d)** $\frac{15}{16}$ **(e)** $\frac{8}{9}$

8.1. **(a)** $9\frac{5}{8}$ **(b)** $15\frac{3}{4}$ **(c)** $11\frac{13}{15}$ **(d)** $8\frac{2}{3}$ **(e)** $19\frac{11}{14}$

9.1. $15\frac{10}{8} = 16\frac{2}{8} = 16\frac{1}{4}$

9.2. **(a)** $7\frac{7}{5} = 8\frac{2}{5}$ **(b)** $12\frac{5}{13}$ **(c)** $20\frac{1}{4}$ **(d)** $15\frac{1}{6}$ **(e)** 18

10.1. **(a)** $6\frac{1}{5}$ **(b)** $16\frac{1}{8}$ **(c)** $8\frac{1}{6}$ **(d)** 11 **(e)** $8\frac{3}{4}$

11.1. Word Problems

1. (3) $2\frac{1}{3}$
$1\frac{2}{3}$
$+\ 3\frac{2}{3}$

$6\frac{5}{3} = 7\frac{2}{3}$

2. (2) $1\frac{3}{4}$
$2\frac{1}{4}$
$+\ 1\frac{1}{4}$

$4\frac{5}{4} = 5\frac{1}{4}$

3. (4) $1\frac{5}{8}$
$\frac{5}{8}$
$+\ 2\frac{7}{8}$

$3\frac{17}{8} = 5\frac{1}{8}$

4. (5) $3\frac{9}{10}$
$+\ \frac{6}{10}$

$3\frac{15}{10} = 4\frac{5}{10} = 4\frac{1}{2}$

5. (3) $\frac{3}{8}$
$+\ \frac{3}{8}$

$\frac{6}{8} = \frac{3}{4}$

FIFTEEN

Subtraction of Like Fractions

1. SUBTRACTING LIKE FRACTIONS

There is ⁴⁄₆ of a cake. You eat ³⁄₆ of the cake. How much is left?

⁴⁄₆ – ³⁄₆

To subtract like fractions, subtract the numerators: 4 – 3 = 1

Then just copy the denominator.
Your answer: ⅙.

You can also use the column form:
Subtract the numerators and bring down the denominator.
Your answer: ⅙.

⁴⁄₆ – ³⁄₆ = ¹⁄

⁴⁄₆ – ³⁄₆ = ⅙

⁴⁄₆
– ³⁄₆
_____
⅙

1.1. Test Yourself

Subtract these fractions: ⁶⁄₇ – ⁴⁄₇

Check your answer. If it is right, go to section 1.2. If it is wrong or you don't know what to do, go to section 1.1A.

1.1A. All you have to do is subtract the numerators. This gives you the numerator of your answer. Then write down the denominator of your two fractions, which is also the denominator of your answer.

Check your answer.

1.2. Practice

Subtract and check by adding.

(a) ¹¹⁄₁₂ – ⁶⁄₁₂

(b) ¾ – ²⁄₄

(c) ⁷⁄₉ – ²⁄₉

(d) ⁸⁄₁₀ – ⁷⁄₁₀

(e) ⅞ – ²⁄₈

Check your answers.

2. REDUCING THE DIFFERENCE TO LOWEST TERMS

Sometimes the difference of two fractions can be reduced to lowest terms.
Example: $\frac{8}{9} - \frac{5}{9}$

Subtract the numerators.

Bring down the denominator.

Your answer: $\frac{3}{9}$.

Reduce it to lowest terms.

Your final answer: $\frac{1}{3}$.

$$
\begin{array}{r}
\frac{8}{9} \\
- \frac{5}{9} \\
\hline
\frac{3}{9} \\
\end{array}
$$

$$\frac{3}{9} = \frac{1}{3}$$

When you reduce your answer to lowest terms, your final answer has a denominator that is different from the denominator of the two fractions you subtracted.

Always give answers to fraction problems in lowest terms.

2.1. Practice

Subtract. Reduce the answers to lowest terms.

(a) $\frac{9}{12}$ (b) $\frac{4}{6}$ (c) $\frac{8}{10}$ (d) $\frac{11}{12} - \frac{3}{12} =$ (e) $\frac{10}{15} - \frac{5}{15} =$

  $- \frac{3}{12}$ $- \frac{2}{6}$ $- \frac{3}{10}$

Check your answers.

3. SUBTRACTING A FRACTION FROM 1

You have 1 gallon of gas. If you put $\frac{3}{4}$ of it into your motor bike, how much gas is left in the can?

$$1 - \frac{3}{4}$$

Put the subtraction in column form: fractions under fractions, units under units (some spaces are empty).

There is nothing you can subtract $\frac{3}{4}$ from. Borrow the 1 and change it to a fraction. $1 = \frac{1}{1}$

Put $\frac{1}{1}$ in the fraction column.

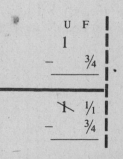

But to subtract you need the same denominator. You must find a fraction equivalent to $\frac{1}{1}$, with 4 as a denominator.

$$\frac{1 \times 4}{1 \times 4} = \frac{4}{4}$$

Put $\frac{4}{4}$ in the fraction column instead of $\frac{1}{1}$.

Now you can subtract (subtract the numerators and bring down the denominator).
Your answer: $\frac{1}{4}$.
There is $\frac{1}{4}$ of a gallon left.
You check it by adding.

After you finish your subtraction, you may have to reduce the answer to lowest terms.

$$1 - \frac{3}{9}$$

Put in column form.

Change 1 to a fraction with 9 as a denominator:

$$1 = \frac{1 \times 9}{1 \times 9} = \frac{9}{9}$$

Subtract $\frac{9}{9} - \frac{3}{9}$ (subtract the numerators and bring down the denominator).
Your answer: $\frac{6}{9}$.

Reduce it to lowest terms.
Your final answer: $\frac{2}{3}$.

3.1. Test Yourself

Do this subtraction: $1 - \frac{5}{8}$.

Check your answer. If it is right, go to section 3.2. If it is wrong or you don't know what to do, go to section 3.1A.

3.1A. Put your subtraction in column form. Be sure to put the fractions under the fractions, and the units under the units.

You have nothing to subtract $\frac{5}{8}$ from. So you must change 1 to a fraction with 8 as a denominator.

$$1 = \frac{1 \times 8}{1 \times 8} = \frac{8}{8}$$

Now you can finish your subtraction. Check it and go to section 3.2.

3.2. Practice

Subtract. Reduce answer to lowest terms if necessary.

(a) $1 - \frac{4}{7}$ **(b)** $1 - \frac{3}{4}$ **(c)** $1 - \frac{4}{10}$

(d) $1 - \frac{8}{12}$ **(e)** $1 - \frac{1}{3}$

Check your answers.

4. SUBTRACTING A FRACTION FROM A WHOLE NUMBER

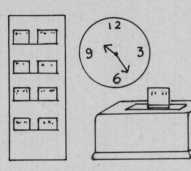

You must work for 3 hours. You have already worked ¼ hour. How long do you still have to work?

$3 - \frac{1}{4}$ hour.

Put in column form.
You have nothing to subtract ¼ from.

So borrow 1 from the units column and change it to a fraction with 4 as a denominator:

$$1 = \frac{1 \times 4}{1 \times 4} = \frac{4}{4}$$

Subtract the fractions column (subtract the numerators and bring down the denominator).

In the units column there is nothing to subtract from 2: just bring it down.

Your answer: 2¾.
You have to work for 2¾ hours.
After you finish your subtraction, you may have to reduce to lowest terms.

	U	F
	3	
−		¼
	2	4/4
	3̸	
−		¼
	2	4/4
−		¼
		¾
	2	4/4
−		¼
	2	¾

4.1. Practice

Subtract. Reduce answers to lowest terms if necessary.

(a) $6 - \frac{3}{8}$ (b) $9 - \frac{5}{7}$ (c) $10 - \frac{4}{5}$ (d) $5 - \frac{2}{3} =$ (e) $17 - \frac{5}{9} =$

Check your answers.

5. SUBTRACTING A MIXED NUMBER FROM A WHOLE NUMBER

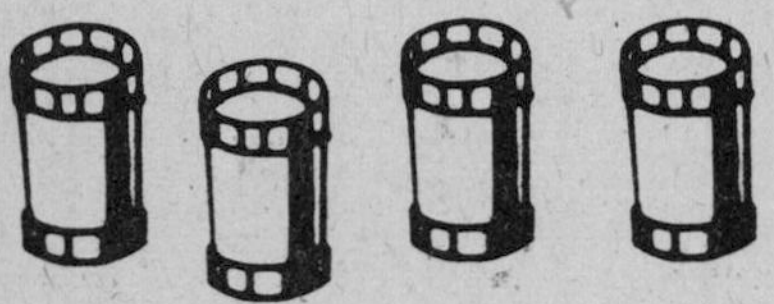

You have 4 rolls of film and you use $2\frac{7}{12}$ rolls. How much film remains?

$$4 - 2\frac{7}{12}$$

Put in column form.
There is nothing you can subtract $\frac{7}{12}$ from.

	U	F
	4	
−	2	$\frac{7}{12}$

So borrow 1 from the units column and change it to a fraction with 12 as a denominator:

$$1 = \frac{1 \times 12}{1 \times 12} \quad \frac{12}{12}$$

	U	F
	3 (4 crossed out)	$\frac{12}{12}$
−	2	$\frac{7}{12}$

Subtract the fraction columns.
Subtract the units column.
Your answer: $1\frac{5}{12}$.
You have $1\frac{5}{12}$ rolls of film left.
You may have to reduce your answer to lowest terms.

	U	F
	3	$\frac{12}{12}$
−	2	$\frac{7}{12}$
	1	$\frac{5}{12}$

Example: $36 - 12\frac{6}{8}$

Put in column form.

	T	U	F
	3	6	
−	1	2	$\frac{6}{8}$

Borrow 1 from the units column.

$$1 = \frac{8}{8}$$

Subtract the fraction column.
Subtract the units column.
Subtract the tens column.
Your answer: $23\frac{6}{8}$.

Reduce to lowest terms.
Your final answer: $23\frac{1}{4}$.

$$\begin{array}{ccc} & 5 & \frac{8}{8} \\ 3 & \cancel{6} & \\ -\ 1 & 2 & \frac{6}{8} \\ \hline 2 & 3 & \frac{2}{8} \end{array}$$

$$23\frac{2}{8} = 23\frac{1}{4}$$

5.1. Practice

(a) 7
  − 5⅖

(b) 12
  − 9³⁄₇

(c) 24
  − 13⁸⁄₁₆

(d) $5 - 2\frac{1}{2} =$

(e) $4 - 1\frac{2}{3} =$

Check your answers.

6. SUBTRACTING MIXED NUMBERS

A farmer picked 3¾ bushels of apples. He sells 2²⁄₄ bushels. How many bushels are left?

$$3\frac{3}{4} - 2\frac{2}{4}$$

Put in column form.
Subtract the fractions column.

Subtract the units column.
Your answer: 1¼.
He has 1¼ bushels left.

You may have to reduce to lowest terms.

$$\begin{array}{cc} 3 & \frac{3}{4} \\ -\ 2 & \frac{2}{4} \\ \hline & \frac{1}{4} \end{array}$$

$$\begin{array}{cc} 3 & \frac{3}{4} \\ -\ 2 & \frac{2}{4} \\ \hline 1 & \frac{1}{4} \end{array}$$

7. SUBTRACTING A FRACTION FROM A MIXED NUMBER

Example: $2\frac{3}{10} - \frac{1}{10}$

$$
\begin{array}{rr}
2 & \frac{3}{10} \\
- & \frac{1}{10} \\
\hline
2 & \frac{2}{10} \\
\end{array}
$$

Your answer: $2\frac{1}{5}$.

$2\frac{2}{10} = 2\frac{1}{5}$

8. SUBTRACTING A WHOLE NUMBER FROM A MIXED NUMBER

Example: $19\frac{2}{3} - 15$

T	U	F
1	9	$\frac{2}{3}$
$-$ 1	5	
0	4	$\frac{2}{3}$

Your answer: $4\frac{2}{3}$

8.1. Practice

(a) $8\frac{9}{12}$　　**(b)** $18\frac{6}{7}$　　**(c)** $7\frac{15}{16}$　　**(d)** $4\frac{9}{15} - 3 =$
　　　$- 6\frac{5}{12}$　　　　$- 9$　　　　$- \frac{7}{16}$　　**(e)** $2\frac{3}{4} - \frac{1}{4} =$

Check your answers. If they are not all correct, read sections 6, 7, and 8 again. Then go to section 9.

9. BORROWING FROM A MIXED NUMBER

A builder ordered $7\frac{2}{5}$ loads of bricks, but he used only $3\frac{4}{5}$ of that amount. How many loads of bricks were not used?

$7\frac{2}{5} - 3\frac{4}{5}$

Put in column form.
Subtract in the fractions column . . . but you can't, because 2 is smaller than 4.

$$
\begin{array}{rr}
7 & \frac{2}{5} \\
- 3 & \frac{4}{5} \\
\hline
\end{array}
$$

Borrow a 1 from the units column and change it to a fraction with 5 as a denominator:

$1 = \frac{5}{5}$

$$
\begin{array}{rr}
6 & \\
\not{7} & \frac{2}{5} + \frac{5}{5} \\
- 3 & \frac{4}{5} \\
\end{array}
$$

You now have $\frac{2}{5} + \frac{5}{5} = \frac{7}{5}$ in the fractions column.
Subtract the fractions, and then subtract the units.
Your answer: $3\frac{3}{5}$ loads of bricks.

$$\begin{array}{rr} 6 & \frac{7}{5} \\ -\ 3 & \frac{4}{5} \\ \hline 3 & \frac{3}{5} \end{array}$$

When you borrow from a mixed number, you must add all you have in the fraction column before you start your subtraction.
After you have finished, be sure to reduce to lowest terms if necessary.

9.1. Test Yourself

Do this subtraction: $35\frac{1}{5} - 22\frac{3}{5}$

Check your answer. If it is right, go to section 10. If it is wrong or you don't know what to do, go to section 9.1A.

9.1A. After you have put your subtraction in column form, you find that you can't subtract the fractions column, because 1 is smaller than 3. So you have to borrow from the units column. You now have $\frac{1}{5} + \frac{5}{5} = \frac{6}{5}$ in the fractions column.
Finish your subtraction and check it. Then go to section 10.

10. BORROWING WHEN SUBTRACTING A FRACTION FROM A MIXED NUMBER

The steps are exactly the same as before.
Example: $22\frac{1}{8} - \frac{5}{8}$

Put in column form.

Subtract in the fractions column . . . but you can't because 1 is smaller than 5.

$$\begin{array}{rrr} 2 & 2 & \frac{1}{8} \\ - & & \frac{5}{8} \\ \hline \end{array}$$

Borrow a 1 from the units column and change it to a fraction with 8 as a denominator.

$$\begin{array}{rrr} & 1 & \\ 2 & \not{2} & \frac{1}{8} + \frac{8}{8} \\ - & & \frac{5}{8} \\ \hline \end{array}$$

You now have $\frac{1}{8} + \frac{8}{8} = \frac{9}{8}$ in the fractions column.

Subtract the fractions.

Subtract the units.

Subtract the tens.

$$\begin{array}{rrr} 2 & 1 & \frac{9}{8} \\ - & & \frac{5}{8} \\ \hline 2 & 1 & \frac{4}{8} \end{array}$$

Reduce your answer.

$$21\frac{4}{8} = 21\frac{1}{2}$$

Your answer: $21\frac{1}{2}$

10.1. Practice

Subtract. Reduce to lowest terms if necessary. You may also have to borrow in other columns (as in a regular subtraction).

(a) $6\frac{1}{4}$ (b) $15\frac{4}{9}$ (c) $25\frac{2}{6}$ (d) $7\frac{3}{5} - 2\frac{4}{5} =$

    $- 2\frac{3}{4}$ $- 10\frac{7}{9}$ $- 23\frac{5}{6}$ (e) $5\frac{2}{7} - 2\frac{4}{7} =$

Check your answers.

11. SUBTRACTION REMINDERS

Make sure that the number you subtract is not larger than the number you subtract from.

Be sure to place the number you subtract under the number you subtract from.

To subtract fractions, mixed numbers, and whole numbers, use the column form. Put fractions under fractions, units under units, tens under tens, and so on.

To subtract $6\frac{7}{8} - 4\frac{3}{8}$ write: $6\frac{7}{8}$

                          $- 4\frac{3}{8}$

To subtract $7 - \frac{3}{4}$ write: 7

                      $- \frac{3}{4}$

You may have empty spaces.

When subtracting fractions, subtract the numerators only. Denominators must be all the same.

If you need to borrow, show all your work.

Reduce to lowest terms when necessary.

You may check by adding.

11.1. More Practice

(a) $87\frac{1}{2}$ (b) $26\frac{4}{10}$ (c) 1 (d) $\frac{37}{50}$

    $- 56$ $- 18\frac{7}{10}$ $- \frac{2}{12}$ $- \frac{12}{50}$

(e) 7 (f) $24\frac{8}{16}$ (g) 1 (h) 28

   $- \frac{1}{8}$ $- \frac{12}{16}$ $- \frac{18}{21}$ $- 14\frac{2}{3}$

Check your answers.

11.2. Word Problems

Before starting this section, review the subtraction word clues in Chapter 4.

1. Sam had $\frac{3}{4}$ of a pound of nails left over from a job. He used $\frac{1}{4}$ of a pound to put up some paneling. What fraction of a pound of nails remained?
(1) $\frac{1}{8}$ **(2)** $\frac{1}{2}$ **(3)** $\frac{1}{3}$ **(4)** $\frac{1}{4}$
(5) none of these

1. 1 2 3 4 5

2. Christian saved $5\frac{1}{4}$ rolls of nickels. He used 2 rolls to buy some gas. How many rolls of nickels did Christian have left?
(1) $3\frac{2}{4}$ **(2)** $2\frac{3}{4}$ **(3)** $2\frac{1}{4}$ **(4)** $3\frac{1}{4}$
(5) none of these

2. 1 2 3 4 5

3. Ernie had a full book of stamps. He paid his bills by mail and used $\frac{2}{3}$ of the book of stamps. What part of the book remained?
(1) $\frac{1}{3}$ **(2)** $1\frac{2}{3}$ **(3)** $1\frac{1}{3}$ **(4)** $\frac{1}{6}$
(5) none of these

3. 1 2 3 4 5

4. Harold plowed $\frac{7}{8}$ of his farm. His brother plowed $\frac{5}{8}$ of his own farm. How much more did Harold plow?
(1) $\frac{1}{8}$ **(2)** $\frac{2}{5}$ **(3)** $\frac{1}{4}$ **(4)** $\frac{1}{2}$
(5) none of these

4. 1 2 3 4 5

5. Jenny bought $6\frac{5}{6}$ yards of fabric. She used $2\frac{2}{6}$ yards for a dress. How many yards of fabric remained?
(1) $4\frac{1}{3}$ **(2)** $4\frac{1}{2}$ **(3)** 4 **(4)** $4\frac{1}{6}$
(5) none of these

5. 1 2 3 4 5

Check your answers. If all your answers are right, begin Chapter 16. If one or two are wrong, make sure that you understand the right answers before starting Chapter 16. If you made more than two mistakes, you should read Chapter 15 again before you start Chapter 16.

12. ANSWERS

1.1. $\frac{2}{7}$

1.2. **(a)** $\frac{5}{12}$ **(b)** $\frac{1}{4}$ **(c)** $\frac{5}{9}$ **(d)** $\frac{1}{10}$ **(e)** $\frac{5}{8}$

2.1. **(a)** $\frac{6}{12} = \frac{1}{2}$ **(b)** $\frac{1}{3}$ **(c)** $\frac{1}{2}$ **(d)** $\frac{2}{3}$ **(e)** $\frac{1}{3}$

3.1.
$$\begin{array}{rcl} 1 & = & \frac{8}{8} \\ -\ \frac{5}{8} & = & \frac{5}{8} \\ \hline & & \frac{3}{8} \end{array}$$

3.2. (a) $\frac{3}{7}$ (b) $\frac{1}{4}$ (c) $\frac{3}{5}$ (d) $\frac{1}{3}$ (e) $\frac{2}{3}$

4.1. (a) $5\frac{5}{8}$ (b) $8\frac{2}{7}$ (c) $9\frac{1}{5}$ (d) $4\frac{1}{3}$ (e) $16\frac{4}{9}$

5.1. (a) $1\frac{3}{5}$ (b) $2\frac{4}{7}$ (c) $10\frac{1}{2}$ (d) $2\frac{1}{2}$ (e) $2\frac{1}{3}$

8.1. (a) $2\frac{1}{3}$ (b) $9\frac{6}{7}$ (c) $7\frac{1}{2}$ (d) $1\frac{3}{5}$ (e) $2\frac{1}{2}$

9.1. $12\frac{3}{5}$

10.1. (a) $6\frac{1}{4} = 5\frac{5}{4}$ (b) $4\frac{2}{3}$ (c) $1\frac{1}{2}$ (d) $4\frac{4}{5}$ (e) $2\frac{5}{7}$
          $-\ 2\frac{3}{4} = 2\frac{3}{4}$
          _____
                $3\frac{2}{4} = 3\frac{1}{2}$

11.1. (a) $31\frac{1}{2}$ (b) $7\frac{7}{10}$ (c) $\frac{5}{6}$ (d) $\frac{1}{2}$ (e) $6\frac{7}{8}$

          (f) $23\frac{3}{4}$ (g) $\frac{1}{7}$ (h) $13\frac{1}{3}$

11.2. Word Problems

    1. (2) $\frac{3}{4}$
          $-\ \frac{1}{4}$
          _____
          $\frac{2}{4} = \frac{1}{2}$

    2. (4) $5\frac{1}{4}$
          $-\ 2$
          _____
          $3\frac{1}{4}$

    3. (1) $1\ \ = \frac{3}{3}$
          $-\ \frac{2}{3} = \frac{2}{3}$
          _____
             $\frac{1}{3}$

    4. (3) $\frac{7}{8}$
          $-\ \frac{5}{8}$
          _____
          $\frac{2}{8} = \frac{1}{4}$

    5. (2) $6\frac{5}{6}$
          $-\ 2\frac{2}{6}$
          _____
          $4\frac{3}{6} = 4\frac{1}{2}$

SIXTEEN

Addition and Subtraction of Unlike Fractions

1. ADDING UNLIKE FRACTIONS

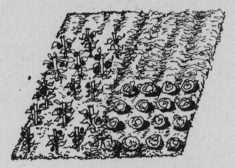 ¼ of the garden was for lettuce. ⅜ was for tomatoes. How much of the garden was for both vegetables?

¼ + ⅜

How can you add these two fractions? You know that you can add fractions only if they have the same denominator.

You can try to replace ¼ by a fraction with 8 as a denominator. For that you can make a graph:

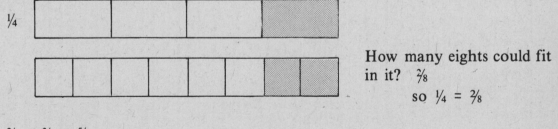

¼

How many eights could fit in it? ²⁄₈

so ¼ = ²⁄₈

²⁄₈ + ⅜ = ⅝

Here is a better way to change unlike fractions to like fractions.
For example: ⅔ + ⅙

⅔: the denominator is 3.
What are the numbers that 3 can divide with no remainder?
3, 6, 9, 12, 15, . . .

167

$\frac{1}{6}$: the denominator is 6.
What are the numbers that 6 can divide with no remainder?
6, 12, 18, 24, 30, . . .

What is the lowest number that they can both divide with no remainder?
3, 6, 9, 12, 15, . . .
6, 12, 18, 24, 30, . . .
It is 6.

6 is the Least Common Denominator (L.C.D.) of both fractions. It is the lowest number that both denominators divide without a remainder.
You replace each fraction by a fraction with 6 as a denominator.

$$\frac{1}{6} = \frac{1}{6} \quad \text{(nothing new to do)}$$

$$\frac{2 \times 2}{3 \times 2} = \frac{4}{6}$$

Now you have two like fractions: you can add them.

$$\frac{1}{6} + \frac{4}{6} = \frac{5}{6}$$

When you need to add two unlike fractions, all you have to do is:

a. Find their Least Common Denominator (L.C.D.).
b. Find fractions equivalent to these fractions with the L.C.D. as denominator (you'll have to raise your first fractions to higher terms).
c. Add the two new fractions.
d. Reduce the answer to lowest terms if necessary, or change it to a whole number or a mixed number if it is an improper fraction.

Here is another example:

$\frac{2}{5} + \frac{3}{4}$

Step a. Numbers that 5 can divide with no remainder:
5, 10, 15, 20, 25, 30, . . .
Numbers that 4 can divide with no remainder:
4, 8, 12, 16, 20, 24, 28, . . .
L.C.D.: 20

Step b. $\frac{2}{5} = \frac{?}{20}$ Answer: $\frac{2}{5} = \frac{8}{20}$
$\frac{3}{4} = \frac{?}{20}$ Answer: $\frac{3}{4} = \frac{15}{20}$

Step c. $\frac{8}{20} + \frac{15}{20} = \frac{23}{20}$

Step d. The answer is an improper fraction. Change it to a mixed number.
$^{23}/_{20} = 1^{3}/_{20}$. Your final answer: $1^{3}/_{20}$.

1.1. Test Yourself

Do this addition: $^{1}/_{7} + ^{2}/_{5}$
Check your answer. If it is right, go to section 2. If it is wrong or you don't know what to do, go to section 1.1A.

1.1A.

Step a. Numbers that 7 can divide with no remainder:

7, 14, 21, 28, 35, 42, 49, . . .

Numbers that 5 can divide with no remainder:

5, 10, 15, 20, 25, 30, 35, 40, 45, . . .

L.C.D.: 35

Step b. Finding the new fractions:

$^{1}/_{7} = ^{?}/_{35}$
$^{2}/_{5} = ^{?}/_{35}$

Now you can finish the problem. Follow the steps carefully.

2. ADDING MORE THAN TWO UNLIKE FRACTIONS

The steps are the same, but you'll have to find the L.C.D. of all the denominators at the same time.

Example: $^{1}/_{6} + ^{2}/_{3} + ^{1}/_{4} + ^{3}/_{5}$

Step a. Numbers that 6 can divide:

6, 12, 18, 24, 30, 36, 42, 48, 54, 60, 66, . . .

Numbers that 3 can divide:

3, 6, 9, 12, 15, 18, 21, 24, 27, 30, 33, 36, 39, 42, 45, 48, 51, 54, 57, 60, 63, 66, . . .

Numbers that 4 can divide:

4, 8, 12, 16, 20, 24, 28, 32, 36, 40, 44, 48, 52, 56, 60, 64, 68 . . .

Numbers that 5 can divide:

5, 10, 15, 20, 25, 30, 35, 40, 45, 50, 55, 60, 65 . . .

L.C.D.: 60

Step b. Finding the new fractions:

$^{1}/_{6} = ^{?}/_{60}$ Answer: $^{1}/_{6} = ^{10}/_{60}$
$^{2}/_{3} = ^{?}/_{60}$ Answer: $^{2}/_{3} = ^{40}/_{60}$
$^{1}/_{4} = ^{?}/_{60}$ Answer: $^{1}/_{4} = ^{15}/_{60}$
$^{3}/_{5} = ^{?}/_{60}$ Answer: $^{3}/_{5} = ^{36}/_{60}$

Step c. Adding the new fractions:

$^{10}/_{60} + ^{40}/_{60} + ^{15}/_{60} + ^{36}/_{60} = ^{101}/_{60}$

Step d. The answer is an improper fraction. It has to be changed to a mixed number:

$^{101}\!/_{60} = 1^{41}\!/_{60}$

2.1. Practice

(a) $\frac{2}{3}$ **(b)** $\frac{3}{5}$ **(c)** $\frac{5}{8}$ **(d)** $\frac{1}{2} + \frac{2}{3} + \frac{3}{4} =$

$\quad + \frac{4}{9}$ $\quad + \frac{4}{15}$ $\quad + \frac{2}{4}$ **(e)** $\frac{5}{8} + \frac{3}{6} + \frac{9}{12} =$

Check your answers.

3. SUBTRACTING UNLIKE FRACTIONS

$\qquad \frac{3}{5} - \frac{1}{3}$

Follow the same steps as for addition.

Step a. $5 \longrightarrow 5, 10, 15, 20, 25, \ldots$

$\qquad 3 \longrightarrow 3, 9, 12, 15, 18, \ldots$

$\qquad$ L.C.D.: 15

Step b. $\frac{3}{5} = ^{9}\!/_{15}$ $\frac{1}{3} = ^{5}\!/_{15}$

Step c. $^{9}\!/_{15} - ^{5}\!/_{15} = ^{4}\!/_{15}$

Step d. The answer doesn't need to be changed. Final answer: $^{4}\!/_{15}$.

3.1. Practice

(a) $\frac{1}{2}$ **(b)** $\frac{2}{3}$ **(c)** $\frac{6}{7}$ **(d)** $\frac{1}{3} - \frac{1}{5} =$

$\quad - \frac{1}{8}$ $\quad - ^{6}\!/_{15}$ $\quad - ^{8}\!/_{14}$ **(e)** $\frac{5}{6} - \frac{6}{9} =$

Check your answers.

4. ADDING UNLIKE MIXED NUMBERS

Unlike mixed numbers are added the same way.

You need to find the L.C.D. of the fractions.

Example: $5^{9}\!/_{12} + 3\frac{5}{8}$

Put the addition in column form. You can't add fractions: they are unlike.

$$\begin{array}{r r} 5 & ^{9}\!/_{12} \\ + \; 3 & \frac{5}{8} \\ \hline \end{array}$$

a. Find L.C.D. of fractions:

12 $\longrightarrow$ 12, 24, 36, 48, 60 . . .

8 $\longrightarrow$ 8, 16, 24, 32, 40, 48 . . .

L.C.D.: 24

b. New fractions: $9/12 = {}^{18}\!/_{24}$ $\quad$ $5/8 = {}^{15}\!/_{24}$

Write your addition again with new fractions.

c. Add the fractions, and then the units.

Your answer: $8{}^{33}\!/_{24}$.

${}^{33}\!/_{24}$ is an improper fraction.

d. Change it to a mixed number: ${}^{33}\!/_{24} = 1{}^{9}\!/_{24}$

Carry the 1 over to the units column.

Your new answer: $9{}^{9}\!/_{24}$

It still needs to be reduced to lowest terms:

Your final answer: $9\tfrac{3}{8}$.

$$\begin{array}{r} 5 \quad {}^{18}\!/_{24} \\ + 3 \quad {}^{15}\!/_{24} \\ \hline 8 \quad {}^{33}\!/_{24} \end{array}$$

$$\begin{array}{r} 1 \\ 5 \quad {}^{18}\!/_{24} \\ + 3 \quad {}^{15}\!/_{24} \\ \hline 8 \quad {}^{33}\!/_{24} \\ 9 \quad {}^{9}\!/_{24} \end{array}$$

$9{}^{9}\!/_{24} = 9\tfrac{3}{8}$

Follow the same steps to add more than two mixed numbers. (Find the L.C.D. of all the denominators.)

4.1. Practice

(a) $13\tfrac{2}{3}$ $\quad$ (b) $24\tfrac{1}{2}$ $\quad$ (c) $25\tfrac{3}{8}$ $\quad$ (d) $42\tfrac{3}{4} + 16{}^{10}\!/_{16} =$

$\quad + 5\tfrac{4}{9}$ $\qquad + 15\tfrac{5}{6}$ $\qquad + 17{}^{5}\!/_{12}$ $\quad$ (e) $223\tfrac{4}{5} + 125{}^{6}\!/_{10} =$

Check your answers.

5. SUBTRACTING UNLIKE MIXED NUMBERS

Always follow the same steps:

$5\tfrac{3}{4} - 2\tfrac{2}{5}$

Put in column form.

You can't subtract the fractions because they are unlike.

a. Find L.C.D.:

4 $\longrightarrow$ 4, 8, 12, 16, 20, 24, 28 . . .

5 $\longrightarrow$ 5, 10, 15, 20, 25, 30 . . . .

L.C.D.: 20

$$\begin{array}{r} 5 \quad \tfrac{3}{4} \\ - 2 \quad \tfrac{2}{5} \\ \hline \end{array}$$

b. New fractions: $\frac{3}{4} = \frac{15}{20}$ $\frac{2}{5} = \frac{8}{20}$

Write your subtraction again with new fractions.

c. Subtract the fractions and then subtract the units.

Your answer: $3\frac{7}{20}$

d. It doesn't need to be changed.

$$\begin{array}{r} 5\ \ ^{15}/_{20} \\ -\ 2\ \ ^{8}/_{20} \\ \hline 3\ \ ^{7}/_{20} \end{array}$$

6. BORROWING

You may have to borrow when subtracting unlike mixed numbers. The borrowing comes after you find the L.C.D.

$5\frac{1}{3} - 3\frac{1}{2}$

Put in column form.

You can't subtract the fractions because they are unlike.

a. Find L.C.D.

$3 \longrightarrow$ 3, 6, 9, 12, 15 . . .

$2 \longrightarrow$ 2, 4, 6, 8, 10, 12 . . .

L.C.D.: 6

b. New fractions: $\frac{1}{3} = \frac{2}{6}$ $\frac{1}{2} = \frac{3}{6}$

Write your subtraction again with new fractions.

You still can't subtract the fractions: $\frac{2}{6}$ is smaller than $\frac{3}{6}$.

Borrow 1 from the units column.

$1 = \frac{1}{1} = \frac{6}{6}$

You now have $\frac{8}{6}$ in the fractions column.

c. Subtract the fractions and then subtract the units.

Your answer: $1\frac{5}{6}$

d. It doesn't need to be changed.

$$\begin{array}{r} 5\ \ \frac{1}{3} \\ -\ 3\ \ \frac{1}{2} \\ \hline \end{array}$$

$$\begin{array}{r} 5\ \ \frac{2}{6} \\ -\ 3\ \ \frac{3}{6} \\ \hline \end{array}$$

$$\begin{array}{r} 4 \\ \cancel{5}\ \ \frac{2}{6}\ +\ \frac{6}{6} \\ -\ 3\ \ \frac{3}{6} \\ \hline \end{array}$$

$$\begin{array}{r} 4\ \ \frac{8}{6} \\ -\ 3\ \ \frac{3}{6} \\ \hline 1\ \ \frac{5}{6} \end{array}$$

6.1. Practice

Subtract. Reduce to lowest terms if necessary. You may have to borrow in several columns (as in a regular subtraction).

(a) $\begin{array}{r} 6\frac{5}{8} \\ -\ 4\frac{1}{4} \\ \hline \end{array}$ (b) $\begin{array}{r} 8\frac{3}{9} \\ -\ 4\frac{2}{3} \\ \hline \end{array}$ (c) $\begin{array}{r} 36\frac{1}{5} \\ -\ 22\frac{7}{15} \\ \hline \end{array}$ (d) $13\frac{4}{10} - 7\frac{3}{5} =$

(e) $2\frac{9}{12} - 1\frac{3}{24} =$

Check your answers.

7. UNLIKE FRACTIONS REMINDERS

To find the L.C.D. of two or several fractions, look for the smallest number all denominators can divide without a remainder.

After finding the L.C.D., replace each fraction by an equivalent fraction having the L.C.D. as its denominator.

Show all your work when you add or subtract.

Reduce answers to lowest terms. Improper fractions must be changed to whole or mixed numbers.

7.1. Word Problems

1. Deborah had $4\frac{1}{2}$ yards of cotton, $2\frac{1}{3}$ yards of silk, and $1\frac{1}{4}$ yards of wool. How much fabric did she have all together?
   (1) $7\frac{11}{12}$ (2) $7\frac{5}{12}$ (3) $8\frac{1}{12}$ (4) $8\frac{10}{12}$
   (5) none of these

   1. 1 2 3 4 5

2. One Tuesday Mr. Maslin ate $\frac{5}{8}$ cup of cottage cheese and $\frac{3}{4}$ cup of spaghetti on his special diet. How much more spaghetti than cottage cheese did he eat?
   (1) $\frac{1}{4}$ (2) 1 (3) $\frac{3}{8}$ (4) $\frac{1}{8}$
   (5) none of these

   2. 1 2 3 4 5

3. Ray unloaded $\frac{1}{2}$ of the truck in 2 hours. Later he unloaded $\frac{1}{3}$ more of the truck. What fraction of the truck was unloaded?
   (1) $\frac{3}{5}$ (2) $\frac{2}{5}$ (3) $\frac{5}{6}$ (4) $\frac{2}{3}$
   (5) none of these

   3. 1 2 3 4 5

4. Pat caught a sunfish that weighed $\frac{3}{4}$ pound and another that weighed $\frac{9}{16}$ pound. How many pounds did the two fish weigh in all?
   (1) $1\frac{5}{16}$ (2) $1\frac{1}{4}$ (3) $1\frac{3}{16}$ (4) $\frac{17}{16}$
   (5) none of these

   4. 1 2 3 4 5

5. Rita's earrings were $1\frac{3}{4}$ inches long. Carmela's were $\frac{7}{8}$ inch long. How much longer are Rita's earrings?
   (1) $\frac{1}{8}$ (2) $\frac{1}{2}$ (3) $\frac{7}{8}$ (4) $\frac{3}{8}$
   (5) none of these

   5. 1 2 3 4 5

6. Howie fed his dog $\frac{1}{2}$ pound of dog food in the morning and $\frac{3}{4}$ of a pound in the evening. How much food did the dog receive all day?
   (1) $1\frac{3}{8}$ (2) $1\frac{1}{6}$ (3) $1\frac{1}{2}$ (4) $1\frac{1}{4}$
   (5) none of these

   6. 1 2 3 4 5

7. Thursday it rained $\frac{3}{4}$ inch. Friday it rained $\frac{3}{8}$ inch. How many inches did it rain in 2 days?
(1) $1\frac{1}{8}$ **(2)** $\frac{7}{8}$ **(3)** $1\frac{3}{4}$ **(4)** 1
(5) none of these

8. A recipe calls for $\frac{3}{4}$ c chopped nuts, $\frac{7}{8}$ c brown sugar, $1\frac{2}{3}$ c of flour and $\frac{1}{2}$ c of water. How many cups of dry ingredients in all?
(1) $3\frac{11}{24}$ **(2)** $3\frac{7}{24}$ **(3)** 3 **(4)** $2\frac{1}{2}$
(5) none of these

Check your answers. If all your answers are right, begin the Evaluation: Fractions. If one or two are wrong, make sure that you understand the right answers before you start the evaluation. If you made more than two mistakes, you should read Chapter 16 again before you start the evaluation.

8. ANSWERS

1.1. $\frac{19}{35}$

2.1. **(a)**
$$\begin{array}{r} \frac{4}{6} \\ +\ \frac{4}{6} \\ \hline \frac{8}{6} = 1\frac{2}{6} = 1\frac{1}{3} \end{array}$$
(b) $\frac{13}{15}$ **(c)** $1\frac{1}{8}$ **(d)** $1\frac{11}{12}$ **(e)** $1\frac{7}{8}$

3.1. **(a)** $\frac{3}{8}$ **(b)** $\frac{4}{15}$ **(c)** $\frac{2}{7}$ **(d)** $\frac{2}{15}$ **(e)** $\frac{1}{6}$

4.1. **(a)**
$$\begin{array}{r} 13\frac{6}{9} \\ +\ 5\frac{4}{9} \\ \hline 18\frac{10}{9} = 19\frac{1}{9} \end{array}$$
(b) $40\frac{1}{3}$ **(c)** $42\frac{19}{24}$ **(d)** $59\frac{3}{8}$ **(e)** $349\frac{2}{5}$

6.1. **(a)** $2\frac{3}{8}$ **(b)** $3\frac{2}{3}$ **(c)** $13\frac{11}{15}$ **(d)** $5\frac{4}{5}$ **(e)** $1\frac{5}{8}$

7.1. Word Problems

1. (3)
$$\begin{array}{r} 4\frac{1}{2} = 4\frac{6}{12} \\ 2\frac{1}{3} = 2\frac{4}{12} \\ +\ 1\frac{1}{4} = 1\frac{3}{12} \\ \hline 7\frac{13}{12} = 8\frac{1}{12} \end{array}$$

2. (4)
$$\begin{array}{r} \frac{3}{4} = \frac{6}{8} \\ -\ \frac{5}{8} = \frac{5}{8} \\ \hline \frac{1}{8} \end{array}$$

3. (3)
$$\begin{array}{r} \frac{1}{3} = \frac{2}{6} \\ +\ \frac{1}{2} = \frac{3}{6} \\ \hline \frac{5}{6} \end{array}$$

4. (1)
$$\begin{array}{r} \frac{3}{4} = \frac{12}{16} \\ +\ \frac{9}{16} = \frac{9}{16} \\ \hline \frac{21}{16} = 1\frac{5}{16} \end{array}$$

5. (3)
$$\begin{array}{r} 1\frac{3}{4} = 1\frac{6}{8} = \frac{14}{8} \\ -\ \frac{7}{8} = \frac{7}{8} = \frac{7}{8} \\ \hline \frac{7}{8} \end{array}$$

6. (4)
$$\begin{array}{r} \frac{1}{2} = \frac{2}{4} \\ +\ \frac{3}{4} = \frac{3}{4} \\ \hline \frac{5}{4} = 1\frac{1}{4} \end{array}$$

7. (1)
$$\begin{array}{r} \frac{3}{4} = \frac{6}{8} \\ +\ \frac{3}{8} = \frac{3}{8} \\ \hline \frac{9}{8} = 1\frac{1}{8} \end{array}$$

8. (2)
$$\begin{array}{r} \frac{3}{4} = \frac{18}{24} \\ \frac{7}{8} = \frac{21}{24} \\ +\ 1\frac{2}{3} = 1\frac{16}{24} \\ \hline 1\frac{55}{24} = 3\frac{7}{24} \\ = 2\frac{11}{12} \end{array}$$

Evaluation: Fractions

Do each problem carefully. Watch the signs.

1. $\frac{7}{6}$
 $\frac{1}{3}$
 $+ \frac{4}{9}$

2. $9\frac{3}{4} \div \frac{13}{16}$

3. $5\frac{3}{7}$
 $9\frac{5}{7}$
 $+ 4\frac{6}{7}$

4. $6\frac{8}{13}$
 $- 5$

5. $6 \times \frac{8}{54}$

6. $\frac{2}{8} \div \frac{1}{2}$

7. $5 \div \frac{25}{26}$

8. 12
 $14\frac{1}{6}$
 $+ 3\frac{1}{4}$

9. $19\frac{5}{8}$
 $- 7\frac{9}{24}$

10. $16\frac{1}{2} \div 7\frac{1}{2}$

11. $10\frac{4}{5} \div \frac{6}{9}$

12. $12\frac{1}{2} \times \frac{8}{10}$

13. $2\frac{1}{7} \div \frac{15}{28}$

14. $\frac{21}{27}$
 $- \frac{6}{27}$

15. $9\frac{15}{30}$
 $- 7\frac{20}{60}$

16. $\frac{6}{9} \times \frac{18}{27} \times \frac{15}{21}$

17. $\frac{15}{18} \times \frac{12}{15}$

18. $10\frac{2}{3}$
 $+ 3\frac{2}{3}$

19. $20\frac{1}{3}$
 $4\frac{4}{7}$
 $+ 2\frac{5}{21}$

20. $8\frac{4}{7} \times \frac{2}{6}$

21. $\frac{8}{9}$
 $+ \frac{3}{4}$

22. $\frac{3}{8} \div \frac{12}{16}$

23. 7
 $- 2\frac{1}{3}$

24. $\frac{11}{12} \div 22$

25. $8\frac{1}{2}$
 $4\frac{3}{4}$
 $+ 9\frac{6}{8}$

26. $16\frac{4}{5}$
 $+ 2\frac{9}{25}$

27. $24\frac{6}{6}$
 $- 15\frac{5}{30}$

28. $\frac{3}{9} \times \frac{18}{30}$

29. $^{15}/_{18} \div 1\frac{1}{3}$

30.
$$
\begin{array}{r}
5^9/_{12} \\
6\frac{2}{3} \\
+\ 3\frac{1}{4} \\
\hline
\end{array}
$$

31. $3^5/_9 \times 9$

32.
$$
\begin{array}{r}
1^1/_{14} \\
-\ ^4/_{14} \\
\hline
\end{array}
$$

33. $^7/_{12} \div 7$

34.
$$
\begin{array}{r}
13^8/_{14} \\
-\ 10^9/_{28} \\
\hline
\end{array}
$$

35. $5\frac{1}{3} \times 2\frac{1}{7}$

36.
$$
\begin{array}{r}
10\frac{3}{4} \\
-\ 4\frac{7}{8} \\
\hline
\end{array}
$$

37. $^{22}/_{24} \times ^4/_{11}$

38. $1^8/_{12} \times 24$

39.
$$
\begin{array}{r}
1 \\
-\ ^{28}/_{32} \\
\hline
\end{array}
$$

40. $^4/_{14} \times ^{35}/_{40}$

41. A childcare center has 2 teachers and 16 children. What is the ratio of teachers to children?
(1) $\frac{1}{8}$　　(2) $^6/_{16}$　　(3) $^2/_8$　　(4) $^8/_{16}$
(5) none of these

41. 1 2 3 4 5

42. Alice brought home nine rolls. Her son ate 3 of them. What fraction of the rolls were gone?
(1) 3　　(2) $6\frac{2}{3}$　　(3) $\frac{1}{3}$　　(4) $\frac{1}{2}$
(5) none of these

42. 1 2 3 4 5

43. A hamburger weighs $\frac{1}{4}$ of a pound. How many pounds will 20 hamburgers weigh?
(1) 6　　(2) 4　　(3) 5　　(4) $5\frac{1}{4}$
(5) none of these

43. 1 2 3 4 5

44. $2\frac{5}{8}$ tons of bricks were on a construction site. $\frac{2}{3}$ of that amount were used. How many tons of bricks were used?
(1) $1\frac{1}{4}$　　(2) $2\frac{1}{8}$　　(3) $1\frac{1}{2}$　　(4) $1\frac{3}{4}$
(5) none of these

44. 1 2 3 4 5

45. 35 feet of rope needs to be cut into pieces only $2\frac{1}{2}$ feet long. How many pieces will there be?
(1) 12　　(2) 13　　(3) 14　　(4) 16
(5) none of these

46. 1 2 3 4 5

46. A bike rider must ride $6\frac{2}{3}$ miles to work. Every $\frac{5}{9}$ of a mile the rider stops to fix the chain. How many stops are there?
(1) 13　　(2) 11　　(3) 10　　(4) 12
(5) none of these

46. 1 2 3 4 5

47. $\frac{3}{4}$ ton of bricks are put onto trucks that can hold $\frac{1}{8}$ ton each. How many trucks are needed?
(1) 6 **(2)** 4 **(3)** 5 **(4)** 8
(5) none of these

47. 1 2 3 4 5

48. You used $\frac{1}{4}$ loaf of bread for French Toast and $\frac{1}{4}$ loaf for sandwiches. How much of the loaf of bread did you use?
(1) $\frac{1}{2}$ **(2)** 1 **(3)** $\frac{3}{4}$ **(4)** $\frac{2}{8}$
(5) none of these

48. 1 2 3 4 5

49. Liz ate $\frac{3}{8}$ of a pizza, James ate $\frac{2}{8}$. How much of the pizza is gone?
(1) $\frac{6}{8}$ **(2)** $\frac{3}{4}$ **(3)** $\frac{1}{2}$ **(4)** $\frac{5}{8}$
(5) none of these

49. 1 2 3 4 5

50. A tanker carried 6 tons of oil and unloaded $3\frac{1}{3}$ tons of it in Seattle, Washington. How many tons of oil were left on the tanker?
(1) $3\frac{1}{3}$ **(2)** 3 **(3)** $2\frac{2}{3}$ **(4)** $2\frac{1}{3}$
(5) none of these

50. 1 2 3 4 5

Check your answers. If you made mistakes, make sure that you understand the right answers. Go back to any chapters that you need to study again.

ANSWERS

1. $1\frac{17}{18}$	**2.** 12	**3.** 20	**4.** $1\frac{8}{13}$
5. $\frac{8}{9}$	**6.** $\frac{1}{2}$	**7.** $5\frac{1}{5}$	**8.** $29\frac{5}{12}$
9. $12\frac{1}{4}$	**10.** $2\frac{1}{5}$	**11.** $16\frac{1}{5}$	**12.** 10
13. 4	**14.** $\frac{5}{9}$	**15.** $2\frac{1}{6}$	**16.** $\frac{20}{63}$
17. $\frac{2}{3}$	**18.** $14\frac{1}{3}$	**19.** $27\frac{1}{7}$	**20.** $2\frac{6}{7}$
21. $1\frac{23}{26}$	**22.** $\frac{1}{2}$	**23.** $4\frac{2}{3}$	**24.** $\frac{1}{24}$
25. 23	**26.** $19\frac{4}{25}$	**27.** $9\frac{1}{2}$	**28.** $\frac{1}{5}$
29. $\frac{5}{8}$	**30.** $15\frac{2}{3}$	**31.** 32	**32.** $\frac{1}{2}$
33. $\frac{1}{12}$	**34.** $3\frac{1}{4}$	**35.** $11\frac{3}{7}$	**36.** $5\frac{7}{8}$
37. $\frac{1}{3}$	**38.** 40	**39.** $\frac{1}{8}$	**40.** $\frac{1}{4}$

41. (1) $\frac{2}{16} = \frac{1}{8}$

42. (3) $\frac{3}{9} = \frac{1}{3}$

43. (3) $\frac{1}{4} \times 20 = \frac{1}{4} \times \frac{\overset{5}{20}}{1} = 5$

44. (4) $\frac{2}{3} \times 2\frac{5}{8} = \frac{2}{\underset{1}{3}} \times \frac{2\frac{1}{1}\frac{7}{8}}{\underset{4}{}} = \frac{7}{4} = 1\frac{3}{4}$

45. (3) $35 \div 2\frac{1}{2} = \frac{\overset{7}{35}}{1} \times \frac{2}{\underset{1}{5}} = 14$

46. (4) $6\frac{2}{3} \div \frac{5}{9} = \frac{\overset{4}{20}}{\underset{1}{3}} \times \frac{\overset{3}{9}}{\underset{1}{5}} = \frac{12}{1} = 12$

47. (1) $\frac{3}{4} \div \frac{1}{8} = \frac{3}{\underset{1}{4}} \times \frac{\overset{2}{8}}{1} = 6$

48. (1) 　$\frac{1}{4}$
　　　$+ \frac{1}{4}$
　　　$\overline{\frac{2}{4}} = \frac{1}{2}$

49. (4) 　$\frac{3}{8}$
　　　$+ \frac{2}{8}$
　　　$\overline{\frac{5}{8}}$

50. (3) 　$6 = 5\frac{3}{3}$
　　　$- 3\frac{1}{3} = 3\frac{1}{3}$
　　　$\overline{ 2\frac{2}{3}}$

SEVENTEEN

Decimals

1. WHAT IS A DECIMAL?

Decimals are fractions that have 10, 100, 1000 . . . for denominators. They can be written either as fractions or as decimals.

$$^6/_{10} = .6$$
$$^{38}/_{100} = .38$$
$$^{259}/_{1000} = .259$$

The DECIMAL POINT shows decimals. Digits to the right of the decimal point are decimals, parts of a whole.

1.1.

The first digit to the right of the decimal point is the TENTH digit. It tells how many tenths are in the number.

tenths
8   .8 = $^8/_{10}$ You say: 8 tenths
5   .5 = $^5/_{10}$ You say: 5 tenths

1.2.

The HUNDREDTHS come just after the tenths, on the right side of the tenths.

tenths	hundredths		
7	3	.73 = $^{73}/_{100}$	You say: 73 hundredths
5	9	.59 = $^{59}/_{100}$	You say: 59 hundredths

1.3.

The THOUSANDTHS come just after the hundredths, on the right side of the hundredths.

tenths	hundredths	thousandths
. 3	5	7	.357 = $357/1000$
			You say: 357 thousandths
. 7	2	9	.729 = $729/1000$
			You say: 729 thousandths

2. READING DECIMALS

When you read decimals, read the number and say in what column you finish.

tenths	hundredths	thousandths	
. 4			.4 You say: 4 tenths
. 3	7		.37 You say: 37 hundredths
. 2	5	6	.256 You say: 256 thousandths

2.1. Practice

Read out loud:

(a) .2   (b) .74   (c) .1   (d) .537   (e) .9

Write the fractions as decimals:

(f) $379/1000$ (g) $5/10$ (h) $19/100$ (i) $725/1000$ (j) $49/100$

Find the decimals and write them with a decimal point:

(k) Ty Cobb's lifetime batting average was 367 thousandths, setting a major league baseball record.

(l) Roger Hornsby holds the record for baseball's highest batting average in a season at 424 thousandths.

Check your answers on page 187. If they are not all right, read sections 1 and 2 again, very carefully. Then go to Section 3.

3. ZERO IS A DIGIT

Look at these fractions and decimals:

tenths	hundredths	
. 0	7	.07 = $7/100$
. 0	3	.03 = $3/100$
. 0	8	.08 = $8/100$

The hundredths column is two places to the right of the decimal point. Put a zero in the tenths column if you only have hundredths.

.5 is 5 tenths; .05 is 5 hundredths

.6 is 6 tenths; .06 is 6 hundredths

Look at these fractions and decimals:

tenths	hundredths	thousandths	
. 0	0	5	$.005 = \frac{5}{1000}$
. 0	3	9	$.039 = \frac{39}{1000}$
. 0	5	6	$.056 = \frac{56}{1000}$

The thousandths column is three places to the right of the decimal point. Put zeros in the columns where you have no other digit.

.5 is 5 tenths   .05 is 5 hundredths   .005 is 5 thousandths

.6 is 6 tenths   .06 is 6 hundredths   .006 is 6 thousandths

3.1. Test Yourself

Write as decimals: $\frac{1}{100}$; 8 thousandths

Check your answers on page 187. If they are right, go to section 3.2. If they are wrong or you don't know what to do, go to section 3.1A.

3.1A.

You can make columns: the tenths column on the right of the decimal point, then the hundredths column, then the thousandths column.

tenths   hundredths   thousandths

Put your digits in the columns. Put zeros in the empty columns to hold the place.

	tenths	hundredths	thousandths
$\frac{1}{100}$ is 1 hundredth	. 0	1	
8 thousandths	. 0	0	8

If you found .1 and .8, it's because you forgot to put the zeros as place holders.

3.2. Practice

Write as decimals:

(a) $\frac{2}{100}$ (b) $\frac{18}{1000}$ (c) $\frac{27}{1000}$ (d) $\frac{9}{100}$

(e) 6 hundredths (f) 7 hundredths (g) 4 thousandths (h) 72 thousandths

Check your answers on page 187.

4. COMPARING DECIMALS

Which is larger: .4 or .04?
Look at the pictures to find out.

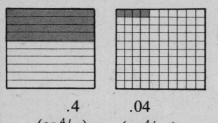

     .4   .04
(or $^4/_{10}$)   (or $^4/_{100}$)

You see that .4 is larger than .04.

Tenths are larger than hundredths.

hundredths are larger than thousandths,

thousandths are larger than ten thousandths, etc.

The places nearer the decimal point are larger than the places farther from it.

Examples: .8 is more than .08
           .86 is more than .086
           .4 is more than .004
           .006 is less than .6
           .045 is less than .45

4.1. Practice

Find the larger decimal:
(a) .25 or .025   **(b)** .034 or .34   **(c)** .007 or .07   **(d)** .1 or .01
(e) .6 or .006
Check your answers on page 187.

5. MIXED DECIMALS

A mixed decimal is made up of a whole number and a decimal.
Example: 5.68 (it is also 5 $^{68}/_{100}$)
The decimal point is between the units and the tenths.

Here are some mixed decimals:

	thousandths	hundreds	tens	units		tenths	hundredths	thousandths
5.68				5	.	6	8	
34.5			3	4	.	5		
679.284		6	7	9	.	2	8	4
4,562.04	4,	5	6	2	.	0	4	

6. READING MIXED DECIMALS

5.68 You say: 5 and 68 hundredths (it is also $5\,^{68}/_{100}$)
34.5 You say: 34 and 5 tenths (it is also $34\,^{5}/_{10}$)
679.284 You say: 679 and 284 thousandths (it is also $679\,^{284}/_{1000}$)
4,562.04 You say: 4,562 and 4 hundredths (it is also $4,562\,^{4}/_{100}$)

AND means DECIMAL POINT. Numbers before AND are whole numbers, numbers after AND are decimals.

6.1. Practice

Read out loud:
(a) 64.5   (b) 19.254   (c) 27.04   (d) 4,348.035
Write as mixed decimals:
(e) $23\,^{4}/_{100}$ (f) $259\,^{27}/_{1000}$ (g) 26 and 8 tenths (h) 329 and 2 hundredths
Find the mixed decimals and write them with a decimal point:
(i) About 88 and 6 tenths inches of snow falls in Buffalo, N.Y., each year.
(j) Micki King, springboard diving champion from the United States, scored 450 and 3 hundredths points in contests during 1972.
(k) Atlanta, Georgia, has about 48 and 66 hundredths inches of rainfall each year.

Check your answers on page 187.

7.

In money, the decimal point is between the dollars and the cents. Cents are hundredths of a dollar.

$\$.01 = \,^{1}/_{100}$ of a dollar. You say: 1 cent, or 1 penny.
$\$.10 = \,^{10}/_{100} = \,^{1}/_{10}$ of a dollar. You say: 10 cents, or 1 dime.
$\$6.78 = 6$ dollars and $^{78}/_{100}$ of a dollar. You say: 6 dollars and 78 cents.

7.1 Practice

Use $ and decimal point to show money.
  9 cents = $.09
  9 dollars and 68 cents = $9.68
(a) 27 cents =
(b) 17 dollars and 5 cents =
(c) 6 cents =
(d) 335 dollars and 9 cents =
(e) 46 dollars and 8 cents =
(f) 4 cents =
(g) 2 cents =
(h) 78 dollars and 35 cents =
(i) 56 dollars and 82 cents =

Check your answers on page 187.

8. ROUNDING OFF

Rounding off decimals reduces the number of decimal places.
Round off .48 to the nearest tenth.
Here is what you do:

Look at the number in the hundredths place.

If that number is under 5, drop it.

If that number is 5 or more, drop it and add 1 to the tenths place.

8 is more than 5.
Add 1 to the tenths place.

Your answer is .5.

.48
.48
.5

Round off .532 to the nearest hundredth.

Look at the number in the thousandths place.

If that number is under 5, drop it.

If that number is 5 or more, drop it and add 1 to the hundredths place.

2 is less than 5.
Drop it.

Your answer is .53

.532
.532
.53

To round off, look at the number to the right of the place you want to round off to.

8.1. Practice

Round off to the nearest tenth: .62▶.6   .78▶.8 **(a)** .4<u>5</u>   **(b)** .6<u>8</u>2   **(c)** .3<u>2</u>8
Round off to the nearest hundredth: .271▶.27 **(d)** .838 **(e)** .386 **(f)** .569 **(g)** .427
Check your answers on page 187.

9. ROUNDING OFF MONEY

Round off $.76 to the nearest dime.

Look at the number of pennies.	
If that number is under 5, drop it.	
If that number is 5 or more, drop it and add 1 to the dimes place.	.7<u>6</u>
6 is more than 5.	.76
Your answer is $.80	

Round off $6.38 to the nearest dollar.

Look at the number of dimes.	$6.<u>3</u>8
If that number is under 5, drop it.	
If that number is 5 or more, drop it and add 1 to the dollars.	$6.38
3 is under 5.	
Drop all the change.	$6.00
Your answer is $6.00	

9.1. Practice

Round off to the nearest dime.
  (a) $.56   **(b)** $.24   **(c)** $5.48
  (d) $2.48 **(e)** $.89
Round off to the nearest dollar.
  (f) $7.38 **(g)** $.62 **(h)** $2.84 **(i)** $6.28 **(j)** $18.43

Check your answers on page 187.

10. DECIMALS REMINDERS

Decimals are fractions which have 10, 100, 1000 . . . for denominators.

The decimal point . means decimals follow to the right.

Tenths come just after the decimal point.
Then come hundredths, then thousandths, etc.

      tenths   hundredths   thousandths

When you read decimals, say the number and say in what column you finish.

    .37 is 37 hundredths
    .052 is 52 thousandths

A mixed decimal is made up of a whole number and a decimal.

The decimal point is between the units column and the tenths column. It is between the whole number and the decimal.

AND means decimal point when you read a mixed decimal out loud.

th means decimal at the end of ten, hundred, thousand, etc.

10 and 3 ten*th*s: 10.3
100 and 42 hundred*th*s: 100.42
1,000 and 225 thousand*th*s: 1,000.225

In money, the decimal point is between the dollars and the cents.
    $.07 = $^7/_{100}$ of a dollar. You say: 7 cents or 7 pennies.

    34 dollars and 78 cents is written $34.78.

Rounding off reduces the number of decimal places.

    .57 rounded off to the nearest tenth is .6.

    .463 rounded off to the nearest hundredth is .46.

11. ANSWERS

1.1

2.1 (a) 2 tenths   (b) 74 hundredths   (c) 1 tenth

     (d) 537 thousandths   (e) 9 tenths

     (f) .379   (g) .5   (h) .19   (i) .725   (j) .49

     (k) .367   (l) .424

3.1 (a) .01   (b) .008

3.2 (a) .02   (b) .018   (c) .027   (d) .09   (e) .06

     (f) .07   (g) .004   (h) .072

4.1 (a) .25   (b) .34   (c) .07   (d) .1   (e) .6

6.1 Say: (a) 64 and 5 tenths   (b) nineteen and 254 thousandths

     (c) 27 and 4 hundredths   (d) 4,348 and 35 thousandths

     (e) 23.04   (f) 259.027   (g) 26.8   (h) 329.02

     (i) 88.6   (j) 450.03   (k) 48.66

7.1 (a) $0.27   (b) $17.05   (c) $0.06   (d) $355.09

     (e) $46.08   (f) $0.04   (g) $.02   (h) $78.35

     (i) $56.82

8.1 (a) .5   (b) .7   (c) .3   (d) .84

     (e) .39   (f) .57   (g) .43

9.1 (a) $.60   (b) $.20   (c) $5.50   (d) $2.50

     (e) $.90   (f) $7.00   (g) $1.00   (h) $3.00

     (i) $6.00   (j) $18.00

EIGHTEEN

Addition of Decimals

1. ADDING DECIMALS

Martha bicycles .2 of a mile to the post office and .7 of a mile to her office. How far does she travel altogether?

.2 + .7 of a mile

To add decimals, use column form.
Put . under .
Put tenths under tenths.

```
    .  tenths
    .    2
  + .    7
  ─────────
    .    2
  + .    7
    .    9
```

Add tenths.
Bring down the decimal point.
Your answer is .9 of a mile.

Each decimal point has to be exactly under the one above.

All you have to do is add as usual. Then when you are finished, you put the decimal point under the others.
If you don't remember how to add regular numbers, look at Chapters 1 and 2 (pages 14–36).

1.1. Test Yourself

Do this addition: .34 + .05
Check your answer on page 197. If it is right, go to section 1.1B. If it is wrong or you don't know what do to, go to section 1.1A.

1.1A.

Write your first number: .34
then put the next decimal point: .

Your numbers will be written this way:

    .34
 + .05
 ‾‾‾‾‾

Now add the hundredths column, then the tenths column.
Put the decimal point exactly under the others.

Check your answer on page 197 and go to section 1.2.

1.2.

Sometimes there is nothing to add in a column.
       Example: .533 + .02 + .14

Use column form.
Put . under .
Put tenths under tenths, hundredths under hundredths, thousandths
under thousandths.

Add the thousandths column. Since there is nothing to add to the
3, just bring it down.

Add the hundredths column.

Add the tenths column. Put your decimal point under the others.
Your answer is .693.

.	tenths	hundredths	thousandths
.	5	3	3
.	0	2	
+ .	1	4	

.	tenths	hundredths	thousandths
.	5	3	3
.	0	2	↓
+ .	1	4	
			3

.	5	3	3
.	0	2	
+ .	1	4	
		9	3

.	5	3	3
.	0	2	
+ .	1	4	
.	6	9	3

1.3. Practice

Remember to put the decimal point in the answer.

(a)	.3	(b)	.16	(c)	.256	(d)	.16	(e)	.54 + .12 =
	+ .5		+ .32		+ .523		.025		
							.304		
							+ .2		

Check your answers on page 197.

2. ADDING MIXED DECIMALS

A jogger ran 3.4 miles on Sunday and 2.3 miles on Wednesday. What was the total distance the jogger ran?

$$3.4 + 2.3 \text{ miles}$$

Use column form.
Put units under units, decimal points under decimal points, tenths under tenths.

Add the tenths column.

Add the units column.
Bring down the decimal point.

Your answer is 5.7. The jogger ran 5.7 miles.

units	.	tenths
3	.	4
+ 2	.	3
3	.	4
+ 2	.	3
		7
3	.	4
+ 2	.	3
5	.	7

2.1. Test Yourself

Do this addition: 7.3 + 2.5
Check your answer on page 197. If it is right, go to section 2.2. If it is wrong or you don't know what to do, go to section 2.1A.

2.1A.

Write your first number: 7.3
then put the next decimal point: .
Your numbers will be written this way:

$$7.3$$
$$+\ 2.5$$

Now all you have to do is add the tenths column, then the units column. When you are finished, put the decimal point exactly under the others. Check your answers on page 197.

2.2. Practice

(a) 1.32   (b) 4.03   (c) 23.5   (d) 36.132   (e) 33.04 + 15.92 =
     + 6.25   + 2.621   + 14.4   + 43.457

Check your answers on page 197.

3. ADDING DECIMALS AND MIXED DECIMALS AT THE SAME TIME

Example: 26.4 + 12.301 + 10.136 + .002

All you have to do is line up your numbers so that all the decimal points are under each other.

Put in column form. Put decimal points under decimal points. Then put tens under tens, units under units, tenths under tenths, hundredths under hundredths, thousandths under thousandths.

Add, starting from the right: add thousandths column, add hundredths column, add tenths column, add units column, add tens column. Bring down the decimal point.

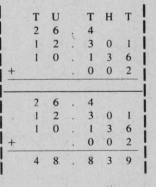

T	U	.	T	H	T
2	6	.	4		
1	2	.	3	0	1
1	0	.	1	3	6
+		.	0	0	2
2	6	.	4		
1	2	.	3	0	1
1	0	.	1	3	6
+		.	0	0	2
4	8	.	8	3	9

Your answer is 48.839.

3.1. Test Yourself

Do this addition:

$$24.36 + .027$$

Check your answer on page 197. If it is right, go to section 3.2. If it is wrong or you don't know what to do, go to section 3.1A.

3.1A.

It is always the same thing. Just remember to put your points one under the other.
Here is how to put your numbers in column form: 24.36
                                                   +   .027

Add each column, starting from the right. Here is what you'll have after you add the thousandths column and the hundredths column:

```
   24.36
+    .027
─────────
      87
```

Finish adding. Bring down the decimal point. Check your answer on page 197.

3.2. Practice

(a)   14.031   (b)   .403   (c)   5.624   (d)   .55
       12.526   .235   12.134   +   .44   (h) 82.634 + 7.053 + .201 =
         .302   +   .151   + 3.021
    + 31.14
                     (e)   .632   (f)   8.004   (g)   .27   (i) 13.041 + 12.603 =
                          +   .24   .542   .11   (j)   .054 +   .723 =
                                          + 1.231   +   .51

Check your answers on page 197. If they are not all right, read sections 1 to 3 again, very carefully. Then go to section 4.

4. CARRYING OVER

Decimal numbers are carried over to the next column just as in regular addition.
Remember to line up your decimal points, and to put the decimal point in your answer.

Everything else is regular adding.
Example: .89 + .57

Use column form. Put decimal points under decimal points, tenths under tenths, hundredths under hundredths.

units	.	tenths	hundredths
	.	8	9
+	.	5	7

Add the hundredths column: 9 + 7 = 16
Carry the 1 from the 16 over to the top of the tenths column.

	.	①	
	.	8	9
+	.	5	7
			6

Add the tenths column: 1 + 8 + 5 = 14
Carry the 1 from the 14 over to the units column.

	.	①	
	.	8	9
+	.	5	7
①		4	6

Bring down the decimal point.

| 1 | . | 4 | 6 |

Your answer is 1.46.

If you don't remember how to carry over in addition, look at sections 4–8 in Chapter 2 (pages 17–21).

4.1. Test Yourself

Do this addition: 26.54 + 7.68
Check your answer on page 197. If it is right, go to section 4.2. If it is wrong or you don't know what to do, go to section 4.1A.

4.1A.

Put your addition in column form:
$$\begin{array}{r} 26.54 \\ +\ 7.68 \\ \hline \end{array}$$

Add the hundredths column. Here is what you get:

$$\begin{array}{r} ① \\ 26.54 \\ +\ 7.68 \\ \hline 2 \end{array}$$

Continue adding. You'll have to carry over again.
When you are finished adding, put the decimal point in your answer.
Check it on page 197.

4.2. Practice

(a)   251.59   (b) 95.8   (c) 19.34   (d) 7.65   (e)   .584 + .625 =
      + 38.8   +   .9   + 76.89   .943
                                          + 2.054

Check your answers on page 197.

5. DOLLARS AND DECIMAL POINTS–ADDING MONEY

A sales clerk rings up $14.95, $7.50, and $3.98 on the cash register. What is the total cost of the items, before tax?

$$\$14.95 + \$7.50 + \$3.98$$

To add money, use the column form.
Put $ on the left of the first number.
Line up the decimal points.
Line up tens, units, tenths, and hundredths.

Do your addition (you need to carry over several times).
Bring down the decimal point.
Bring down the dollar sign.
Your answer is $26.43
(26 dollars and 43 cents).

```
$ 1 4 . 9 5
    7 . 5 0
+   3 . 9 8

    ⑪⑪ ①
$ 1 4 . 9 5
    7 . 5 0
+   3 . 9 8

$ 2 6 . 4 3
```

5.1. Practice

Remember to bring down the decimal point and the dollar sign.
(a) $5.48   (b) $27.50   (c) $ .38
      + 2.69   + 48.95   + .47
(d) $10.99 + $3.51 =   (e) $25.50 + $25.50 =

Check your answers on page 197.

6. ADDITION REMINDERS

Be sure to put the numbers you are adding in the correct column before you add.

2.34 + .56 + 26.4 + 5.098   To add, write in column form:

$$\begin{array}{r} 2.34 \\ .56 \\ 26.4 \\ +\quad 5.098 \\ \hline \end{array}$$

Put decimal points under decimal points.
Line up tens, units, tenths, hundredths, thousandths, etc.
Start adding from the right side.
Carry over if you need to.
Bring down the decimal point.
To add money, put $ on the left side of the top number and of the answer.

7. MORE PRACTICE

(a)
$$\begin{array}{r} .43 \\ +\ .64 \\ \hline \end{array}$$

(b)
$$\begin{array}{r} .982 \\ .341 \\ +\ .684 \\ \hline \end{array}$$

(c)
$$\begin{array}{r} 5.8 \\ 7.032 \\ +\ 5.42 \\ \hline \end{array}$$

(d)
$$\begin{array}{r} 26.1 \\ 38.53 \\ +\ 5.004 \\ \hline \end{array}$$

(e)
$$\begin{array}{r} \$26.57 \\ .98 \\ +\ .32 \\ \hline \end{array}$$

(f)
$$\begin{array}{r} \$109.00 \\ +\ 37.58 \\ \hline \end{array}$$

(g)
$$\begin{array}{r} \$29.99 \\ +\ 4.98 \\ \hline \end{array}$$

(h)
$$\begin{array}{r} \$\ .56 \\ .23 \\ +\ .79 \\ \hline \end{array}$$

Check your answers on page 197.

8. WORD PROBLEMS

Before starting, look at section 9 of Chapter 2 (pages 21–22). That will remind you of the ADDITION WORD CLUES.

1. It is .5 of a mile from home to the bank. From the bank to the cleaner's it is .3 of a mile. How far is it from home to the cleaner's by way of the bank?
   (1) .08   **(2)** .8   **(3)** .9   **(4)** 1
   (5) NT

2. Consuela's temperature rose 3.6 degrees when she was sick. Her normal temperature is 98.6 degrees. What was her temperature when she was sick?
   (1) 98.96   **(2)** 103.6   **(3)** 102.2   **(4)** 101.6
   (5) NT

3. A bettor won $26.40 on a horse race at Brandywine Track. The Exacta paid the same person $62.80. How much was won on the two races together?
   (1) $99.30   **(2)** $88.20   **(3)** $98.48   **(4)** $89.20
   (5) NT

4. What is the total of the following distances:
   12.5 miles, 68.8 miles, 3.2 miles, and 250.4 miles?
   (1) 334.9   **(2)** 354.8   **(3)** 344.9   **(4)** 334.09
   (5) NT

5. Mail-order instructions for building a cabinet cost $2.25 plus $.75 for postage and handling. How much does it all amount to?
   (1) $2.95   **(2)** $3.05   **(3)** $3.00   **(4)** $3.10
   (5) NT

Check your answers on page 197. If all your problems are right, start Chapter 19. If one or two are wrong, make sure you understand the right answers, and start Chapter 19. If you made more than two mistakes, it's better that you read Chapter 18 again. PRACTICE. Then start Chapter 19.

9. ANSWERS

1.1. .39

1.3. (a) .8   **(b)** .48   **(c)** .779   **(d)** .689   **(e)** .66

2.1. 9.8

2.2. (a) 7.57   **(b)** 6.651   **(c)** 37.9   **(d)** 79.589   **(e)** 48.96

3.1. 24.387

3.2. (a) 57.999   **(b)** .789   **(c)** 20.779   **(d)** .99   **(e)** .872

　　 (f) 9.777   **(g)** .89   **(h)** 89.888   **(i)** 25.644   **(j)** .777

4.1. 34.22

4.2. (a) 290.39   **(b)** 96.7   **(c)** 96.23   **(d)** 10.647   **(e)** 1.209

5.1. (a) $8.17   **(b)** $76.45   **(c)** $ .85   **(d)** $14.50   **(e)** $51.00

7.   (a) 1.07   *(b)* 2.007   **(c)** 18.252   **(d)** 69.634   **(e)** $27.87

　　 (f) $146.58   **(g)** $34.97   **(h)** $1.58

8.   WORD PROBLEMS

1. (2)　　.5　　　　　　**4. (1)** 12.5
　　　　+ .3　　　　　　　　　68.8
　　　　―――　　　　　　　　　3.2
　　　　　.8　　　　　　　　+ 250.4
　　　　　　　　　　　　　　――――
2. (3)　98.6　　　　　　　334.9
　　　　+　3.6　　　　**5. (3)** $2.25
　　　　――――　　　　　　+　.75
　　　　102.2　　　　　　　　――――
3. (4) $26.40　　　　　　　$3.00
　　+　62.80
　　　――――
　　　$89.20

NINETEEN

Subtraction of Decimals

1. SUBTRACTING DECIMALS

.9 of the town's telephones didn't work due to the storm. By noon, .5 of the phones were repaired. What part was still out of service?

.9 − .5

To subtract decimals, use column form.

Put . under .

Put tenths under tenths.

Subtract tenths.

Bring down the decimal point.

Your answer: .4 of the town's telephones were still out of service.

Each decimal point has to be exactly under the one before.

	. tenths
	9
−	. 5

	. 9
−	. 5
	. 4

1.1.

Here is an example: .376 − .243

Use column form.

Put . under .

Put tenths under tenths, hundredths under hundredths, thousandths under thousandths.

Subtract thousandths,

subtract hundredths,

subtract tenths.

tenths	hundredths	thousandths
. 3	7	6
− . 2	4	3
		3

tenths	hundredths	thousandths
. 3	7	6
− . 2	4	3
	3	3

tenths	hundredths	thousandths
. 3	7	6
− . 2	4	3
. 1	3	3

Bring down the decimal point.
Your answer is .133.

Check by adding.

```
 . 1   3   3
+ . 2   4   3
─────────────
 . 3   7   6
```

Just subtract as usual. Then when you are finished, bring down the decimal point.
If you don't remember how to subtract regular numbers, look at Chapter 4 (pages 37–48).

1.2. Test Yourself

Do this subtraction: .58 − .32
Check your answer on page 209. If it is right, go to section 1.3. If it is wrong or you don't
know what to do, go to section 1.2A.

1.2 A.

It is most important to line up your numbers. Each decimal point has to be exactly under the
one before.
Write your first number: .58
Then put the next decimal point: .
Your numbers will be written this way: .58
                                          − .32

Then subtract as usual, starting from the right. When you are finished, bring down the
decimal point. Check by adding. Check your answer on page 209.

1.3. Practice

Remember to put the decimal point in the answer.

(a)   .675   (b)   .42   (c)   .88   (d)   .97   (e) .9 − .6 =
      − .434   − .21   − .26   − .52

Check your answers on page 209.

1.4.

Sometimes there is nothing to subtract in a column.
Example: .553 − .42

Use column form.
Line up the decimal points.
Line up tenths, line up hundredths, line up thousandths.

Subtract the thousandths column.
Since there is nothing to subtract from 3, just bring it down.

Subtract the hundredths column.
Subtract the tenths column.
Bring down the decimal point.
Your answer is .133.

Check by adding.

tenths	hundredths	thousandths
.5	5	3
− .4	2	

tenths	hundredths	thousandths
.5	5	3
− .4	2	
		3

.5	5	3
− .4	2	
.1	3	3

.1	3	3
+ .4	2	
.5	5	3

1.5. Practice

Remember to put the decimal point in the answer.

(a)   .758   (b)   .389   (c)   .58   (d) .956 − .82 =
   − .24   − .17   − .4   (e) .682 − .53 =

Check your answers on page 209.

2. BORROWING

You can borrow when you subtract decimals, as in regular subtraction.
Example: .873 − .549

Use column form.
Line up the decimal points.
Line up tenths, line up hundredths, line up thousandths.

Subtract the thousandths column . . . but you can't because 3 is less than 9.

So borrow 1 from the hundredths column.
You now have 13 thousandths.
Subtract the thousandths column:
13 − 9 = 4

Subtract the hundredths column.
Subtract the tenths column.

Bring down the decimal point.
Your answer is .324

Check.

tenths	hundredths	thousandths
. 8	7	3
− . 5	4	9
	6	13
. 8	⁊	⁊
− . 5	4	9
		4
	6	13
. 8	⁊	⁊
− . 5	4	9
. 3	2	4
	①	
. 3	2	4
+ . 5	4	9
. 8	7	3

If you don't remember how to borrow in subtraction, look at sections 1–7.2 of Chapter 5 (pages 49–57).

2.1. Test Yourself

this subtraction: .85 − .37

 your answer on page 209. If it is right, go to section 2.2. If it is wrong or you don't
 what to do, go to section 2.1A.

2.1A.

Put your subtraction in column form:
$$\begin{array}{r} .85 \\ -\ .37 \\ \hline \end{array}$$

You can't subtract hundredths because 5 is less than 7. So borrow 1 from the tenths column.

$$\begin{array}{r} 7\ \ 15 \\ .\ 8\ \ 5 \\ -\ .\ 3\ \ 7 \\ \hline \end{array}$$

Now you can finish your subtraction.
Remember to bring down the decimal point. Check by adding.

2.2.

You may have to borrow more than once. Here is an example:
$$.342 - .175$$

Put in column form.

Subtract the thousandths column . . . but 2 is less than 5.

Borrow 1 from the hundredths column.
You now have 12 thousandths.
Subtract the thousandths column.
Subtract the hundredths column . . . but 3 is less than 7.

Borrow 1 from the hundredths column.
Subtract the hundredths column.
Subtract the tenths column.
Bring down the decimal point.

Your answer is .167.

Check.

tenths	hundredths	thousandths
. 3	4	2
− . 1	7	5
	3	12
. 3	4̸	2̸
− . 1	7	5
		7
2	13	
	3̸	12
. 3̸	4̸	2̸
− . 1	7	5
. 1	6	7
①	①	
. 1	6	7
+ . 1	7	5
. 3	4	2

2.3. Practice

(a)   .284 (b)   .571 (c)   .63 (d)   .58 (e) .913 − .085 =
    − .127   − .259   − .29   − .09

Check your answers on page 209.

3. SUBTRACTING MIXED DECIMALS

$$4.5 - 1.3$$

Use column form.
Put units under units, decimal points under decimal points, tenths under tenths.

units	.	tenths
4	.	5
− 1	.	3

4	.	5
− 1	.	3
		2

Subtract the tenths column.
Subtract the units column.
Bring down the decimal point.
Your answer is 3.2.

4	.	5
− 1	.	3
3	.	2

Check.

3	.	2
+ 1	.	3
4	.	5

3.1.

When you subtract mixed decimals, you may also have to borrow.

The average temperature in Miami in March is 71.3 degrees. In January, the average temperature is 67.5 degrees. How much warmer is it in March than in January?
   71.3 − 67.5 degrees

Use column form.
Subtract the tenths column . . . but 3 is less than 5.

T	U	.	Tth
7	1	.	3
− 6	7	.	5

Borrow 1 from the units column.
Subtract the tenths column.
Subtract the units column . . . but 0 is less than 7.

	0	13	
7	1̶	.	3̶
− 6	7	.	5
			8

Borrow 1 from the tens column.
Subtract the units column.
Subtract the tens column.
Bring down the decimal point.
Your answer is 3.8.

It is 3.8 degrees warmer in March than in January.
Check.

```
        6 10
         0    13
        7 1 . 7
    −  6 7 . 5
       0 3 . 8
    ───────────
        ① ①
        3 . 8
    +  6 7 . 5
      7 1 . 3
```

3.2 Practice

(a)   48.24   **(b)**   5.5   **(c)**   3.19   **(d)**   7.524   **(e)** 64.2 − 38.5 =
    − 13.57   − 1.8   − 1.34   − 2.896

Check your answers on page 209.

4. SUBTRACTING A DECIMAL FROM A MIXED DECIMAL

Example: 237.25 − .75
Just line up your numbers so that all the decimal points are under each other.

Put in column form.
Subtract hundredths.
Subtract tenths . . . but 2 is less than 7.

Borrow 1 from the units column.
Subtract tenths.
Subtract units: just bring down the 6.
Subtract tens: just bring down the 3.
Subtract hundreds: just bring down the 2.
Bring down the decimal point.
Your answer is 236.50.

Check.

```
    H T U . Tth Hth
    2 3 7 .  2   5
    −     .  7   5
    ─────────────────
                 0

            6    12
    2 3 7 .  2   5
    −       .  7   5
    2 3 6 .  5   0
    ─────────────────
          ①
    2 3 6 .  5   0
    +       .  7   5
    2 3 7    2   5
```

4.1 Practice

(a)　5.48　(b)　764.34　(c)　18.12　(d)　35.54　(e) 46.52 − .27 =
− .25　　−　.27　　−　.76　　− 27.88

Check your answers on page 209.

5.　DOLLARS AND DECIMAL POINTS–SUBTRACTING MONEY

The insurance company paid $23.25 of a $46.79 bill. How much of the bill was the client's responsibility?

$46.79 − $23.25

To subtract money, use column form.
Put $ on the left of the first number.
Line up the decimal points.
Line up tens, units, tenths, hundredths.

Subtract pennies, subtract dimes, subtract dollars.

Bring down the decimal point.
Bring down the dollar sign.
Your answer:$23.54 was the client's responsibility.

Check.

```
$ 46.79
− 23.25
─────────
$ 46.79
− 23.25
$ 23  54
─────────
$ 46.79
− 23.25
$ 23.54
─────────
$ 23.54
+ 23.25
$ 46.79
```

5.1.

You may need to borrow when you subtract money.

How much do you save if you buy hamburger on sale for $.98 a pound instead of $1.19 a pound?

$$\$1.19 - \$.98$$

Put in column form.
Subtract pennies.
Subtract dimes . . . but 1 is less than 9.

Borrow 1 from the units column.
Subtract dimes.
Subtract dollars (there is nothing left).

Bring down the decimal point.
Bring down the dollar sign.
Your answer is $.21. You save $.21 a pound.

Check.

```
        $ 1 . 1 9
      -     . 9 8
                1

          0  11
        $ 1 . 1 9
      -     . 9 8
              2 1

          0  11
        $ 1 . 1 9
      -     . 9 8
        $   . 2 1

        $   . 2 1
      +     . 9 8
        $ 1 . 1 9
```

5.1.A

The shoe repair bill comes to $2.39 with the tax. You pay with a five dollar bill. How much change do you get?

$$\$5.00 - \$2.39$$

Put in column form.
Subtract pennies . . . but 0 is less than 9.
You must borrow from the dimes column, but there is a zero. You can't borrow from zero.

Borrow from 50 instead: $50 - 1 = 49$

Subtract pennies, subtract dimes, subtract dollars.
Bring down the decimal point.
Bring down the dollar sign.

```
        $ 5 . 0 0
      - 2 . 3 9

          4  9 10
        $ 5 . 0 0
      - 2 . 3 9
        $ 2 . 6 1

        $ 2 . 6 1
      + 2 . 3 9
        $ 5 . 0 0
```

Your answer is $2.61. You get that much as change.

Check.

5.2 Practice

Put the decimal point and the dollar sign in your answer.

(a) $ 7.86 (b) $ .89 (c) $ 24.95 (d) $8.00 − $3.49 =
   − 2.53   − .46   − 12.68 (e) $43.90 − $14.55 =

Check your answers on page 209.

6. SUBTRACTION REMINDERS

Make sure that the number you subtract is not larger than the number you subtract from.
Be sure to put the number you subtract under the number you subtract from:

     56.75 − .29   To subtract, write in column form:

$$\begin{array}{r} 56.75 \\ - \quad .29 \\ \hline \end{array}$$

Put decimal points under decimal points.
Line up tens, units, tenths, hundredths, thousandths, etc.
Start subtracting from the right side.

Borrow if you need to. Show all your work when you borrow.

Leave some space on top of your subtraction in case you need to borrow several times.

Bring down the decimal point.

To subtract money, put $ on the left side of the top number and of the answer.

Check all subtraction by adding.

7. MORE PRACTICE

(a)   .985   (b)   3.6   (c)   8.02   (d)   .632
     − .453   − .4   − 5.48   − .457

(e)   23.96   (f)   5.86   (g)   20.05   (h)   .384
     − 8.97   − .32   − 14.35   − .195

Check your answers on page 209.

8. WORD PROBLEMS

Before starting, look at section 7 of Chapter 4 (pages 45–46). That will remind you of the SUBTRACTION WORD CLUES.

1. Monthly bills for 750 kilowatt hours of electricity are $37.25 in the winter and $42.12 in the summer. What is the difference between the summer and winter rates?
   (1) $4.23   (2) $4.87   (3) $5.13   (4) $5.00
   (5) NT

2. A set of 8 glasses is priced at $7.95. If you have a coupon the price $2.00 less. How much is the set with the coupon?
   (1) $5.95   (2) $5.00   (3) $2.00   (4) $2.95
   (5) NT

3. A report says most women live to be 79.4 years old and most men live to be 68.5 years. How many years more do women outlive men?
   (1) 10.9   (2) 19   (3) 10.09   (4) 9.9
   (5) NT

4. It was 101.3 degrees outside. Overnight the temperature dropped 16.8 degrees. What was the morning temperature?
   (1) 85.1   (2) 94.5   (3) 84.5   (4) 80.1
   (5) NT

5. A haircut, shampoo, and set is $10.50 at Louise's Beauty Salon. The price is $12.25 at Maxine's Salon. What is the savings if a customer goes to Louise's?
   (1) $2.50   (2) $.75   (3) $2.00   (4) $2.75
   (5) NT

6. Coffee dropped in price from $2.60 a pound to $2.39. How much was the price reduced?
  (1) $.39   **(2)** $.49   **(3)** $2.29   **(4)** $.31
  (5) NT

7. 4.6 million people received questionnaires in the mail. .7 of the people did not answer the questions. How many people filled out the questionnaire?
  (1) 5.13 million   **(2)** 3.9 million   **(3)** 4 million
  (4) 3.7 million   **(5)** NT

8. A large construction company has assets of $8.4 million and debts of $9.9 million. The company has how much more in debts?
  (1) $1.13 million **(2)** $1.5 million **(3)** $15 million
  (4) $18.3 million **(5)** NT

Check your answers on page 210. If all your problems are right, start Chapter 20. If one or two are wrong, make sure you understand the right answers, and start Chapter 20. If you made more than two mistakes, it's better that you read Chapter 19 again. Then start Chapter 20.

9. ANSWERS

1.2 .26

1.3 **(a)** .241   **(b)** .21   **(c)** .62   **(d)** .45   **(e)** .3

1.5 **(a)** .518   **(b)** .219   **(c)** .18   **(d)** .136   **(e)** .152

2.1 .48

2.3 **(a)** .157   **(b)** .312   **(c)** .34   **(d)** .49   **(e)** .828

3.2 **(a)** 34.67 **(b)** 3.7   **(c)** 1.85   **(d)** 4.628 **(e)** 25.7

4.1 **(a)** 5.23   **(b)** 764.07 **(c)** 17.36 **(d)** 7.66 **(e)** 46.25

5.2 **(a)** $5.33 **(b)** $.43   **(c)** $12.27 **(d)** $4.51 **(e)** $29.35

7. **(a)** .532   **(b)** 3.2   **(c)** 2.54   **(d)** .175   **(e)** 14.99

   (f) 5.54   **(g)** 5.70   **(h)** .189

8. WORD PROBLEMS

1. (2) $42.12
      − 37.25
      $ 4.87

2. (1) $7.95
      − 2.00
      $5.95

3. (1) 79.4
      − 68.5
      10.9

4. (3) 101.3
      − 16.8
      84.5

5. (5) $12.25
      − 10.50
      $ 1.75

6. (4) $2.60
      − 2.29
      $0.31

7. (2) 4.6
      − .7
      3.9

8. (2) $9.9
      − 8.4
      $1.5

TWENTY

Multiplication of Decimals

1. MULTIPLYING DECIMALS

Example: $.64 \times .2$

Put in column form.
But the decimal points do not need to be under each other.

Multiply. Don't worry about the decimal points.

Count the digits to the right of the decimal points. Altogether, there are 3 digits to the right of the points.

So there must be 3 digits to the right of point in the answer.
Start from the right. Count 3 digits. Put your point.

Your answer is .128.
This is the same answer: 0.128.

```
  .6 4
×   .2
```

```
  .6 4
×   .2
  1 2 8
```

```
  .64
×  .2
  .1 2 8
```

1.1.

Here is another example:
     $.6 \times .8$

Put in column form.
Multiply.

```
   . 6
× . 8
   4 8
```

How many digits to the right of the points altogether? 2.
Start from the right and count 2 digits.

211

Put your decimal point.

Your answer is .48.

This is the same answer: 0.48

```
  . 6
× . 8
  . 4 8
```

Be careful: in multiplication, decimal points don't always go under each other like in addition or subtraction.

Multiply like regular numbers. Then find the place of the decimal point in the answer.

If you don't remember how to multiply regular numbers, look at Chapters 6 and 7 (pages 59–78).

1.2. Test Yourself

Do this multiplication: .57 × .3

Check your answer on page 220. If it is right, go to section 1.3. If it is wrong or if you don't know what to do, go to section 1.2A.

1.2A.

Multiply your numbers as if they were regular numbers.

```
    57
×    3
   171
```

Then count how many digits altogether are on the right of decimal points:

```
   .57
×   .3
   171
```

There are 3 digits to the right of decimal points. So in your answer you must also have 3 digits to the right of the decimal point.

Put the decimal point in its place and check your answer on page 220.

1.3. Practice

(a) .36   (b) .452   (c)   .63   (d) .989 × .4 =
     × .3   × .5   × .24   (e) .32 × .44 =

Check your answers on page 220.

2. MULTIPLYING DECIMALS, WHOLE NUMBERS, AND MIXED DECIMALS

You can multiply any kind of numbers. The steps are always the same.
Example: multiplying a mixed decimal by a whole number.
$$3.42 \times 4$$

Put in column form.

$$\begin{array}{r} 3.42 \\ \times\ \ 4 \end{array}$$

Multiply as regular numbers.
Count decimal places. This problem has 2.

$$\begin{array}{r} 3.42 \\ \times\ \ 4 \\ \hline 13\ 68 \end{array}$$

Start at the right. Count 2 places to the left.
Put the decimal point between 3 and 6.

$$\begin{array}{r} 3.42 \\ \times\ \ 4 \\ \hline 13.68 \end{array}$$

Your answer is 13.68.

Count the decimal places in the problem. There must be the same number of decimal places in the answer.
Example: multiplying a mixed decimal by a decimal.
$$3.82 \times .3$$

Put in column form. Multiply as regular numbers.
Count decimal places. There are 3 altogether.

$$\begin{array}{r} 3.82 \\ \times\ \ .3 \\ \hline 11\ 46 \end{array}$$

Start from the right and count 3 places.
Put the decimal point.

$$\begin{array}{r} 3.82 \\ \times\ \ .3 \\ \hline 1.146 \end{array}$$

Your answer is 1.146.

2.1. Practice

Don't forget to put the decimal point in the answer.
(a) .4 × .4   (b) .74 × .2 (c) 4.8 × 6   (d) 8.234 × .2 (e) 6.3 × 3

Check your answers on page 220.

3. Using Zeros

When you count the total number of decimal places, you may need to use zeros to make up the right amount.
Example: .3 × .3

Put in column form. Multiply as regular numbers.

$$\begin{array}{r} .3 \\ \times\ .3 \\ \hline 9 \end{array}$$

Count the decimal places. There are 2 altogether.
But in your answer you have only 1 digit.
Put a zero to make up the second digit.

$$\begin{array}{r} .3 \\ \times\ .3 \\ \hline 09 \end{array}$$

Now you can put your decimal point.

$$\begin{array}{r} .3 \\ \times\ .3 \\ \hline .09 \end{array}$$

Your answer is .09.

Put in zeros to make decimals in the answer the same as decimals in the problem.
Here is another example: .04 × .2

Put in column form and multiply.

$$\begin{array}{r} .04 \\ \times\ .2 \\ \hline 08 \end{array}$$

Count the decimal places. There are 3 altogether.
So you need a zero to make 3 digits in your answer.

$$\begin{array}{r} .04 \\ \times\ .2 \\ \hline 008 \end{array}$$

Now you can put your decimal point.

$$\begin{array}{r} .04 \\ \times\ .2 \\ \hline .008 \end{array}$$

Your answer is .008.

3.1. Test Yourself

Do this multiplication: .43 × .2
Check your answer on page 220. If it is right, go to section 3.2. If it is wrong or you don't know what to do, go to section 3.1A.

3.1A.

Put your multiplication in column form and multiply as you do for regular numbers:

```
  43
×  2
────
  86
```

Then count the decimal places:

```
 .43
× .2
────
  86
```

There are 3 decimal places altogether, but you only have 2 digits in your answer. You need a zero to make up for the third digit. Put the zero. Put the decimal point.
Check your answer on page 220.

3.2. Practice

(a) .2 × .02   (b) .41 × .2   (c) .03 × .3   (d) .03 × .2   (e) .32 × .3

Check your answers on page 220.

4. DECIMAL POINTS AND LONGER MULTIPLICATIONS

Multiply as if the numbers were regular numbers. Then put the decimal point in its place.
Example: 432.5 × .35

Put in column form.
Multiply. Don't worry about the decimal points.
Count decimal places in the problem. There are 3.

```
  432.5
×   .35
```

```
  432.5
×   .35
──────
  21625
 129750
──────
 151375
```

Put the decimal point three places from the right.
Your answer is 151.375.

```
  432.5
×   .35
──────
  21625
 129750
──────
 151.375
```

If you don't remember how to do long multiplication, look at sections 1–6 of Chapter 7 (pages 71–76).

4.1. Practice

(a)   3.24   (b)   5.6   (c)   43.017   (d) 5.7 × 2.4 =
     × .15   × 2.7   ×   .25   (e) 8.24 × .13 =

Check your answers on page 220.

4.2.

You may need to add zeros to make up the right number of decimal places in the answer.
Example: .206 × .24

Put in column form. Multiply. Don't worry about the decimal
points.
Count the decimal places in the problem. There are 5.

You need a zero to make up 5 digits in the answer.
Put the zero. Put the decimal point.
Your answer is .04944.

```
    .206
  ×  .24
    824
   4120
   4944

    .206
  ×  .24
    824
   4120
  .04944
```

4.3. Practice

You may need more than one zero.

(a)   .482   (b)   .347   (c)   .04   (d)   .256   (e) .456 × .079 =
     × .206   × .12   × .14   × .15

Check your answers on page 220.

5. DECIMAL POINTS, DOLLARS—MULTIPLYING MONEY

The steps are the same for multiplying money. Put $ to the left of the first number, and in the final answer.

A union carpenter in Alaska earns $16.85 per hour. What is the salary of an 8-hour day, before taxes?

$16.85 × 8

Put in column form. Multiply.
Count the decimal places: 2.

Put the decimal point 2 places from the right.
Bring down the dollar sign.
The salary for 8 hours is $134.80

$$\begin{array}{r} \$\ 16.85 \\ \times\quad 8 \\ \hline 13480 \end{array}$$

$$\begin{array}{r} \$\ 16.85 \\ \times\quad 8 \\ \hline 134.80 \end{array}$$

5.1. Practice

Don't forget to bring down the dollar sign.

(a) $24.75
   × 3

(b) $38.95
   × 6

(c) $1.99
   × 5

(d) $20.84 × 2 =

(e) $342.27 × 3 =

Check your answers on page 220.

6. MULTIPLICATION REMINDERS

The decimal points do not need to be in a column.

3.45 × .2

To multiply, write in column form:

$$\begin{array}{r} 3.54 \\ \times \quad .2 \\ \hline \end{array}$$

Multiply as usual. Don't worry about decimal points until the final answer.

The number of decimal places in the problem tells you how many decimal places there are in the answer.

$$\begin{array}{r} 3.54 \\ \times \quad .2 \\ \hline .708 \end{array}$$ three decimal places

three decimal places in the answer.

You may need to add zeros to make up the number of decimal places:

$$\begin{array}{r} .3 \\ \times .2 \\ \hline .06 \end{array}$$

To multiply money, put $ on the left side of the top number and of the answer.

7. MORE PRACTICE

(a) $\begin{array}{r} .06 \\ \times .29 \\ \hline \end{array}$ (b) $\begin{array}{r} .413 \\ \times .2 \\ \hline \end{array}$ (c) $\begin{array}{r} .576 \\ \times .004 \\ \hline \end{array}$ (d) $\begin{array}{r} .593 \\ \times 7 \\ \hline \end{array}$

(e) $\begin{array}{r} .06 \\ \times 24 \\ \hline \end{array}$ (f) $\begin{array}{r} .895 \\ \times 382 \\ \hline \end{array}$ (g) $\begin{array}{r} 17.4 \\ \times .20 \\ \hline \end{array}$ (h) $\begin{array}{r} 48.03 \\ \times .16 \\ \hline \end{array}$

Check your answers on page 220.

8. WORD PROBLEMS

Before starting, look at section 10 of Chapter 6 (page 68). That will remind you of the MULTIPLICATION WORD CLUES.

1. An eyebrow pencil weighs .07 ounces. How many ounces would 6 pencils weigh?
   (1) 4.2   (2) 42   (3) .42   (4) .48
   (5) NT

2. A half-gallon bottle of laundry detergent is $2.29. What is the price of 4 bottles?
   (1) $9.16   (2) $9.38   (3) $8.30   (4) $9.61
   (5) NT

3. Tomatoes are $.69 a pound. How much will 3.5 pounds cost? (Round off to the nearest penny.)
   (1) $2.45   (2) $2.40   (3) $2.41   (4) $2.42
   (5) NT

4. A store has 43.5 dozen felt-tipped markers in stock. How many markers are in stock? There are 12 in 1 dozen.
   (1) 532   (2) 522   (3) 512   (4) 402
   (5) NT

5. A movie ticket costs $2.75. What is the price for 5 tickets?
   (1) $13.00   (2) $13.85   (3) $13.75   (4) $12.75
   (5) NT

6. A job at a lead storage battery plant pays $6.50 per hour. What would you earn (before taxes) in one day if you work 7.5 hours? (Round off to the nearest penny.)
   (1) $48.75   (2) $48.50   (3) $47.58   (4) $48.50
   (5) NT

7. Ramona ate .5 pounds of fish every day for 7.5 days. How many pounds of fish did she eat in all?
   (1) 38.5   (2) 3.75   (3) 4.5   (4) 3.85
   (5) NT

8. Gasoline is $.62 a gallon. What is the cost of 10.4 gallons?
(Round off to the nearest penny.)
(1) $6.48   **(2)** $6.54   **(3)** $6.45   **(4)** $6.84
(5) NT

Check your answers on page 221. If all your problems are right, start Chapter 21. If one or two are wrong, make sure you understand the right answers, and start Chapter 21. If you made more than two mistakes, it's better that you read Chapter 20 again. Then start Chapter 21.

9. ANSWERS

1.2. .171

1.3. (a) .108   **(b)** 2.260   **(c)** .1512   **(d)** .3956   **(e)** .1408

2.1. (a) .16   **(b)** .148   **(c)** 28.8   **(d)** 1.6468   **(e)** 18.9

3.1. .086

3.2. (a) .004   **(b)** .082   **(c)** .009   **(d)** .006   **(e)** .096

4.1. (a) .486   **(b)** 15.12   **(c)** 10.75425   **(d)** 13.68   **(e)** 1.0712

4.3. (a) .099292 **(b)** .04164   **(c)** .0056   **(d)** .0384   **(e)** .036024

5.1. (a) $74.25   **(b)** $233.70 **(c)** $29.85   **(d)** $41.68 **(e)** $1,026.87

7. (a) 0.0174   **(b)** 0.0826   **(c)** .002304   **(d)** 4.151   **(e)** 1.44

   (f) 341.89   **(g)** 3.48   **(h)** 7.6848

8. WORD PROBLEMS

1. (3)
$$
\begin{array}{r}
.07 \\
\times\ 6 \\
\hline
.42
\end{array}
$$

7. (2)
$$
\begin{array}{r}
7.5 \\
\times\ .5 \\
\hline
37.5
\end{array}
$$

2. (1)
$$
\begin{array}{r}
\$\ 2.29 \\
\times\ \ \ \ 4 \\
\hline
\$\ 9.16
\end{array}
$$

8. (3)
$$
\begin{array}{r}
10.4 \\
\times\ \$.62 \\
\hline
208 \\
6240 \\
\hline
\$\ 6.448 \\
\$\ 6.45
\end{array}
$$

3. (4)
$$
\begin{array}{r}
.69 \\
\times\ 3.5 \\
\hline
345 \\
2070 \\
\hline
\$2.415 \\
\$2.42
\end{array}
$$

4. (2)
$$
\begin{array}{r}
43.5 \\
\times\ \ 12 \\
\hline
870 \\
4350 \\
\hline
522.0
\end{array}
$$

5. (3)
$$
\begin{array}{r}
\$\ 2.75 \\
\times\ \ \ \ 5 \\
\hline
\$13.75
\end{array}
$$

6. (1)
$$
\begin{array}{r}
\$\ 6.50 \\
\times\ \ \ 7.5 \\
\hline
3250 \\
45500 \\
\hline
\$48.750 \\
\$48.75
\end{array}
$$

TWENTY–ONE

Division of Decimals

1. DIVIDING A DECIMAL BY A WHOLE NUMBER

The steps are the same as for regular division, but you have to remember to put the decimal point.
Example: .6 ÷ 2
Use the other division sign.

Put the decimal point right on top of the other decimal point. Do it before you start dividing, so you don't forget it.

Divide as usual.

Your answer is .3.

Check by multiplying.

If you don't remember how to divide regular numbers, look at Chapter 8 (pages 79–94).

$$2\overline{)\,.6}$$

$$2\overline{)\,\overset{.}{.6}}$$

$$2\overline{)\,\overset{.3}{.6}}$$

$$\begin{array}{r} .3 \\ \times\ 2 \\ \hline .6 \end{array}$$

1.1. Test Yourself

Do this division: $4\overline{)\,.8}$
Check your answer on page 238. If it is right, go to section 1.2. If it is wrong or you don't know what to do, go to section 1.1A.

1.1A.

Before you start, put the decimal point right on top of the other decimal point:

$$4\overline{)\ .8}$$

Now, all you have to do is divide as usual. Do your division and check it by multiplying.

1.2. Practice

Divide and check by multiplying.

(a) $3\overline{)\ .9}$ (b) $2\overline{)\ .8}$ (c) $3\overline{)\ .6}$ (d) $.75 \div 5$

                                                          (e) $.84 \div 7$

Check your answers on page 238.

2. ZERO IS A DIGIT

After you put your decimal point, you cannot always start dividing. Put zeros in your answer until you can start dividing.

Example: $.246 \div 6$

$$6\overline{)\ .246}$$

Use the other division sign.
Put the decimal point.

$$6\overline{)\ .\overset{.0}{246}}$$

Start dividing. 6 doesn't go into 2. Put a zero on top of the 2.

$$6\overline{)\ \overset{.04}{.246}}$$

Join the first two columns. 6 goes into 24 four times.

$$6\overline{)\ \overset{.041}{.246}}$$

Finish dividing as usual. Your answer is .041.

Check.

$$\begin{array}{r} .041 \\ \times\ 6 \\ \hline .246 \end{array}$$

2.1.

You may need to put more than one zero.

Example: .245 ÷ 35
Use the other division sign. Put the decimal point.

Start dividing. 35 doesn't go into 2. Put a zero on top of the 2 because it goes zero times.

Join the first two columns. 35 goes zero times into 24. Put a zero on top of the 4.

Join the first three columns. 35 goes into 245 seven times.
Your answer is .007

Check.

$$35\overline{).245}$$

$$\overset{.0}{35\overline{).245}}$$

$$\overset{.00}{35\overline{).245}}$$

$$\overset{.007}{35\overline{).245}}$$

$$\begin{array}{r} .007 \\ \times\ 35 \\ \hline 035 \\ 0210 \\ \hline 0.245 = .245 \end{array}$$

2.2. Test Yourself

Do this division: $6\overline{).546}$
Check your answer on page 238. If it is right, go to section 2.3. If it is wrong or you don't know what to do, go to section 2.2A.

2.2A.

Put your decimal point:

$$6\overline{).546}$$

6 goes into 5 zero times. Put a zero on top of the 5.

$$\overset{.0}{6\overline{).546}}$$

Join the first two columns. Now you can finish this division.
Check by multiplying.

2.3. Practice

Divide and check by multiplying.

(a) 4 $\overline{)\,.168}$ **(b)** 21 $\overline{)\,.84}$ **(c)** 3 $\overline{)\,.189}$ **(d)** .564 ÷ 6 =

(e) .172 ÷ 43 =

Check your answer on page 238.

3. DIVIDING A MIXED DECIMAL BY A WHOLE NUMBER

The steps are the same. Just put the decimal point when you arrive at it.
Example: 25.65 ÷ 5

Use the other division sign. Start your division as usual.
5 doesn't go into 2. Join the first two columns.
5 goes into 25 five times.

You have come to the decimal point. Put it in your answer before
you continue.
It must be right on top of the other decimal point.

Finish your division as usual.
Your answer is 5.13.

$$\begin{array}{r} 5 \\ 5\,\overline{)\,25.65} \end{array}$$

$$\begin{array}{r} 5. \\ 5\,\overline{)\,25.65} \end{array}$$

$$\begin{array}{r} 5.13 \\ 5\,\overline{)\,25.65} \end{array}$$

$$\begin{array}{r} 5.13 \\ \times\ \ 5 \\ \hline 25.65 \end{array}$$

Check by multiplying.

The decimal point is like a STOP signal. You must put it in your answer before you
continue your division.

3.1. Test Yourself

Do this division: 12.84 ÷ 6
Check your answer on page 238. If it is right, go to section 3.2. If it is wrong or you don't
know what to do, go to section 3.1A.

3.1A.

Use the other division sign, and start your division as usual:

$$\begin{array}{r} 2 \\ 6\,\overline{\smash)12.84} \end{array}$$

STOP! You've come to the decimal point. Put it in your answer:

$$\begin{array}{r} 2. \\ 6\,\overline{\smash)12.84} \end{array}$$

This decimal point must be exactly on top of the other one.
Finish your division as usual, and check it by multiplying.

3.2

You may have to put the decimal point before you start to divide.
Example: $5.67 \div 7$

Use the other division sign. Start as usual.
7 doesn't go into 5, so join the first two columns.
But STOP! You must put your decimal point now because your
answer will be on top of the 6.

Divide as usual. Your answer is .81.

Check by multiplying.

$$7\,\overline{\smash)5.67}$$

$$7\,\overline{\smash)5.67}$$

$$\begin{array}{r} .81 \\ 7\,\overline{\smash)5.67} \end{array}$$

$$\begin{array}{r} .81 \\ \times\ \ 7 \\ \hline 5.67 \end{array}$$

3.3. Practice

(a) $7\,\overline{\smash)14.84}$ (b) $9\,\overline{\smash)9.927}$ (c) $3\,\overline{\smash)7.65}$ (d) $3.69 \div 9 =$

(e) $5.46 \div 6 =$

Check your answers on page 238.

4. CHANGING A WHOLE NUMBER TO A MIXED DECIMAL

Putting zeros after the decimal point does not change the value of a number.
Look:

$$6 = 6.0 = 6.00 = 6.000 = 6.0000$$

They all mean 6.
Think of 6 dollars: you can write it $6 or $6.00. It's always six bucks.

$$37 = 37.0 = 37.00 = 37.000, \text{ etc.}$$

Don't forget to put the decimal point! 370 is NOT equal to 37!

4.1. Test Yourself

Change 15 to a mixed decimal.
Check your answer on page 238. If it is right, go to section 5. If it is wrong or you don't know what to do, go to section 4.1A.

4.1A.

To change 15 to a mixed decimal, put a decimal point after the 5 and add as many zeros as you want:

$$15 = 15.0 = 15.00 = 15.000, \text{ etc.}$$

If you answered 150 or 1500, be careful. You must put the decimal point before you add the zeros.

5. DIVIDING A NUMBER BY A LARGER NUMBER

Can you divide 2 by 4? Yes, if you change 2 to a mixed decimal.

$$4\,\overline{)2}$$

Change 2 to a mixed decimal.

Put 2.0 instead of 2 under the division sign.

4 doesn't go into 2, so join the first two columns.
But STOP! You must put your decimal point now.

Divide as usual. Your answer is .5.

Check.

$$2 = 2.0$$

$$4\,\overline{)2.0}$$

$$4\,\overline{)2.0}^{\,.}$$

$$4\,\overline{)2.0}^{\,.5}$$

$$\begin{array}{r} .5 \\ \times\,4 \\ \hline 2.0 = 2 \end{array}$$

5.1. Test Yourself

Do this division: $5\,\overline{)4.0}$
Check your answer on page 238. If it is right, go to section 5.2. If it is wrong or you don't know what to do, go to section 5.1A.

5.1A.

Change 4 to a mixed decimal: $4 = 4.0$
Put 4.0 instead of 4 under the division sign: $5\,\overline{)4.0}$
5 doesn't go into 4, so join the first two columns. But STOP! Put your decimal point:

$$5\,\overline{)4.0}^{\,.}$$

Finish the division and check it by multiplying.

5.2.

You may need to add more than one zero.

Example: $50 \overline{)2}$

Change 2 to a mixed decimal.

Write the division in new form.
50 doesn't go into 2, so join the first two columns.
Before dividing, put the decimal point.

50 goes into 20 zero times. Put a zero on top of the other zero.

To be able to start dividing, you need one more zero after the 2.
Join the first three columns.

50 goes into 200 four times.

Your answer is .04.

Check.

$$2 = 2.0$$

$$50 \overset{\cdot}{\overline{)2.0}}$$

$$50 \overset{.0}{\overline{)2.0}}$$

$$50 \overset{.0}{\overline{)2.00}}$$

$$50 \overset{.04}{\overline{)2.00}}$$

$$\begin{array}{r} .04 \\ \times\ 50 \\ \hline 00 \\ 200 \\ \hline 2.00 = 2 \end{array}$$

5.3. Practice

(a) $5\overline{)1}$ (b) $8\overline{)4}$ (c) $70\overline{)7}$ (d) $5\overline{)4}$ (e) $6 \div 60 =$

Check your answers on page 238.

5.4.

If you get a remainder when dividing a decimal, a whole number, or a mixed decimal, add zeros and keep dividing until there is nothing left.
Example: $2.7 \div 5$

Put in usual form.
Put the decimal point and join the first two columns.

$$5 \overline{)2.7}$$

5 goes into 27 five times.
$5 \times 5 = 25$. Remainder: 2.

$$\begin{array}{r} .5 \\ 5 \overline{)2.7} \\ -25 \\ \hline 2 \end{array}$$

Add a zero.
Join the remainder and zero.
5 goes into 20 four times.
$5 \times 4 = 20$. There is no remainder.

$$\begin{array}{r} .54 \\ 5 \overline{)2.7} \\ -25 \\ \hline 20 \\ -20 \\ \hline 0 \end{array}$$

Your answer is .54.

Check.

$$\begin{array}{r} .54 \\ \times\ 5 \\ \hline 2.70 = 2.7 \end{array}$$

5.5. Test Yourself

Do this division: $6 \overline{)2.1}$
Check your answer on page 238. If it is right, go to section 5.6. If it is wrong or you don't know what to do, go to section 5.5A.

5.5A.

Put the decimal point and join the first two columns:

6 goes into 21 three times.
$6 \times 3 = 18$. Remainder: 3.

$$6\overline{)2.1}$$

Add a zero to the remainder.
Join the remainder and zero.

$$\begin{array}{r} .3 \\ 6\overline{)2.1} \\ -1\ 8 \\ \hline 30 \end{array}$$

Finish the division and check by multiplying.

5.6. Sometimes you need to add more than one zero.

Example: $4\overline{)1.7}$

$$\begin{array}{r} .4 \\ 4\overline{)1.7} \\ -1\ 6 \\ \hline 10 \end{array}$$

Put the decimal point and join the first two columns.
4 goes into 16 four times. Remainder: 1.
Add a zero. Join the remainder and zero.

$$\begin{array}{r} .42 \\ 4\overline{)1.7} \\ -1\ 6 \\ \hline 10 \\ -\ 8 \\ \hline 20 \end{array}$$

4 goes into 10 two times. Remainder: 2.
Add a zero and join the remainder and zero.

4 goes into 20 five times. There is no remainder.

$$\begin{array}{r} .425 \\ 4\overline{)1.7} \\ -1\ 6 \\ \hline 10 \\ -\ 8 \\ \hline 20 \\ -\ 20 \\ \hline 0 \end{array}$$

Your answer is .425.

Check.

$$\begin{array}{r} .425 \\ \times\quad 4 \\ \hline 1.700 = 1.7 \end{array}$$

5.7. Practice

For some divisions, you also need to put zeros before you can start dividing.
(a) $5\overline{).36}$ **(b)** $10\overline{).99}$ **(c)** $6\overline{).51}$ **(d)** $20\overline{).19}$ **(e)** $1.2 \div 8 =$

Check your answers on page 238.

6. ROUNDING OFF YOUR ANSWER

If you divide more than three places after the decimal point and still have a remainder, round off your answer to three places.
Example: $8.2 \div 7$

```
      1
  7 )8.2
    - 7
      1
```

Put in usual form.
Start dividing.

```
      1.
  7 )8.2
    - 7
      1
```

STOP! Put the decimal point.

```
      1.1
  7 )8.2
    - 7
      1 2
    - 7
      50
```

Continue dividing. There is a remainder.
Add a zero.

```
      1.17
  7 )8.2
    - 7
      1 2
    - 7
      50
    - 49
      10
```

Continue dividing.
There is a remainder. Add a zero.

```
      1.171
  7 )8.2
    - 7
      1 2
    - 7
      50
    - 49
      10
    - 7
      30
```

Continue dividing.
There is a remainder. Add a zero.

```
      1.1714
  7 )8.2
    - 7
      1 2
    - 7
      50
    - 49
      10
    - 7
      30
    - 28
       2
```

Continue dividing.
There is still a remainder.
You can't continue forever.

Round off to three places after the decimal point.
Look at the fourth place. It is less than 5.
Drop the last digit.
Your answer is 1.171.

```
1.171 ④
1.171
```

6.1. Practice

Round off to three places:

(a) 3) 4.57   **(b)** 7).89   **(c)** 6).275   **(d)** 9)9.7   **(e)** 2 ÷ 6 =

Check your answers on page 239.

7. DIVIDING BY A DECIMAL NUMBER

You cannot divide by a decimal number. You must change your division first.
Example: 2.75 ÷ .5

Put in usual form.
Move the decimal point one place to the right.
.5 becomes 5. (that is, 5)

```
.5 ) 2.75

.5. ) 2.75
```

To keep the same value, move the decimal point under the)
one place to the right also.

```
5 ) 2.7.5

5 ) 27.5
```

Your division is now:

Do it as usual.

```
      5.5
5 ) 27.5
   − 25
      2 5
    − 2 5
        0
```

Your answer is 5.5.

Check.

```
    5.5
  × .5
   2.75
```

7.1. Test Yourself

Divide: .8) 3.28

Check your answer on page 239. If it is right, go to section 8. If it is wrong or you don't
know what to do, go to section 7.1A.

7.1A.

You can't divide by a decimal.
Move both decimal points over one place to the right: .8. $\overline{)\ 3.2.8}$
Your division is now: 8 $\overline{)\ 32.8}$
Now finish the problem.
Check it by multiplying.

8. DIVIDING BY A MIXED DECIMAL

You can't divide by a mixed decimal. You must change your division.
Example: 4.8 ÷ 1.2

Put in usual form.

You can't divide by a mixed decimal.
Move the decimal point one place to the right.

Do the same thing under $\overline{)}$.

Your division is now:

Do it as usual. Your answer is 4.

Check.

```
1.2 )4.8

1.2. )4.8

12 )4.8.

12 )48

      4
12 )48

    1.2
   × 4
   ────
    4.8
```

8.1. Practice

(a) .5 $\overline{).565}$ **(b)** 3.6 $\overline{).72}$ **(c)** .7 $\overline{).4991}$ **(d)** 1.944 ÷ .24 =
                                                          (e) .999 ÷ .333 =

Check your answers on page 239..

9. ADDING ZEROS

You need to add zeros if the number you divide doesn't have enough places after the decimal point.
Example: 8 ÷ 1.6

Put in usual form.

You can't divide by a decimal.
Move the decimal point one place to the right.
1.6 becomes 16.

$1.6\overline{)8}$

$1.\overset{\smile}{6.}\overline{)8}$

Do the same under $\overline{)}$.
You need to put a zero to show the new decimal place.
8 becomes 80.

$16\overline{)8.\overset{\smile}{0.}}$

$16\overline{)80}$

Your division is now:

$16\overline{)\overset{5}{80}}$

Do it as usual. Your answer is 5.

$$\begin{array}{r} 5 \\ \times\ 1.6 \\ \hline 8.0 = 8 \end{array}$$

Check.

9.1. Test Yourself

Divide: $.15\overline{)34.5}$
Check your answer on page 239. If it is right go to section 9.2. If it is wrong or you don't know what to do, go to section 9.1A.

9.1A.

You can't divide by a decimal. Move both decimal points over two places to the right.
$\quad.\overset{\smile}{15.}\overline{)34.\overset{\smile}{50.}}$
$\qquad$ You need a zero to show the new decimal place.
Your division is now: $15\overline{)3450}$
Do it and check by multiplying.

9.2. Practice

Round off your answer if necessary.
(a) $.08\overline{)56.8}$ **(b)** $.3\overline{)1}$ **(c)** $.6\overline{)2}$ **(d)** $.7\overline{)2}$ **(e)** $.05\overline{)42.6}$

Check your answers on page 239.

10. DOLLARS AND DECIMAL POINTS—DIVIDING MONEY

The rules are the same. Don't forget the dollar sign.
The gas pump reads $5.67 for 9 gallons of gas. How much is each gallon?

$5.67 ÷ 9

$$9\overline{)\$5.67}$$

Put in usual form. Put $ next to the number you divide.

$$\begin{array}{r} \$\ .63 \\ 9\overline{)\$5.67} \\ -5\ 4 \\ \hline 27 \\ -\ 27 \\ \hline 0 \end{array}$$

Divide as usual. Put $ next to answer.
Each gallon costs $.63.

Remember: the numbers to the left of the decimal point are dollars. The numbers to the right of the decimal point are cents.

10.1. Practice

(a) $2\overline{)\$12.48}$ (b) $5\overline{)\$382.35}$ (c) $25\overline{)\$75.25}$ (d) $289.40 ÷ 4 =$
                                                            (e) $434.62 ÷ 62 =$

Check your answers on page 239.

11. DIVISION REMINDERS

If division problems are in this form: $2.4 ÷ 1.2 =$

be sure to put them in this form: $1.2\overline{)2.4}$

You may also use the form $2.4\left|\dfrac{1.2}{}\right.$ if you already know it.

Put the decimal point in the answer when you arrive at it.

$$\begin{array}{cc} 5 & \\ 5\overline{)25.5} \end{array} \quad \text{STOP!} \quad \begin{array}{cc} 5. & \\ 5\overline{)25.5} \end{array}$$
Put the decimal point.

The decimal point in the answer must be right on top of the other one.

You can't divide by a decimal or by a mixed decimal.

Move the decimal point to the right until you get a whole number

outside the $\overline{)}$. Do the same thing inside the $\overline{)}$.

$$2.7\overline{)1.35} \qquad \text{becomes} \qquad 27\overline{)13.5}$$

Add zeros if you need to.

$$1.53\overline{)2.8}\qquad\text{becomes}\qquad 153\overline{)280}$$

Add zeros if you want to get rid of the remainder.

$$
\begin{array}{r}
.675 \\
4\overline{)2.7} \\
-\underline{2\,4} \\
30 \\
-\underline{28} \\
20 \\
-\underline{20} \\
0
\end{array}
$$

If you still get remainders, round off your answer to three decimal places.

When you divide money, don't forget the dollar sign.

12. WORD PROBLEMS

Before starting, look at section 12 of Chapter 8. That will remind you of the DIVISION WORD CLUES.

1. You spend $10.54 to fill your gas tank. Gas is $.68 a gallon. How many gallons did you buy?
   (1) 15   **(2)** 15.5   **(3)** 16.6   **(4)** 16.5
   (5) NT

2. A jet flew 1,994.3 miles in 5.5 hours. About how many miles did the jet fly in 1 hour?
   (1) 362.4   **(2)** 362.6   **(3)** 362   **(4)** 365.2
   (5) NT

3. It costs $.15 to mail a letter. How many stamps can you buy with $1.65?
   (1) 11   **(2)** 12   **(3)** 15   **(4)** 13
   (5) NT

4. 10 mini-boxes of candy weighed a total of 5 pounds. What part of a pound did each mini-box weigh?
   (1) 5.5   **(2)** 1.5   **(3)** .5   **(4)** .51
   (5) NT

5. Gerald spends $62.40 a year to buy the *Daily News* 6 days a week. There are 52 weeks in a year. How much does Gerald spend each week?
   (1) $.20   (2) $12.20   (3) $.12   (4) $0.120
   (5) NT

6. Number 10 business envelopes cost $6.40 per box of 500. How many boxes can be purchased with $38.40?
   (1) 5   (2) 7   (3) 8   (4) 6
   (5) NT

7. 3 pounds of tomatoes cost $.99. What is the price of 1 pound?
   (1) $0.35   (2) $ .34   (3) $ .33   (4) .36
   (5) NT

8. A nylon jacket is $8.95. How many jackets can be purchased for $35.80?
   (1) 4   (2) 5   (3) 6   (4) 3
   (5) NT

Check your answers on page 239. If all your problems are right, start Chapter 22. If one or two are wrong, make sure you understand the right answers, and start Chapter 22. If you made more than two mistakes, read Chapter 21 again. Then start Chapter 22.

13. ANSWERS

1.1. .2

1.2. (a) .3   **(b)** .4   **(c)** .2   **(d)** .15   **(e)** .12

2.2. .091

2.3. (a) .042   **(b)** .04   **(c)** .063   **(d)** .094   **(e)** .004

3.1. 2.14

3.3. (a) 2.12   **(b)** 1.103   **(c)** 2.55   **(d)** .41   **(e)** .91

4.1. 15.0

5.1. .8

5.3. (a) .2   **(b)** .5   **(c)** .1   **(d)** .8   **(e)** .1

5.5. .35

5.7. (a) .072   **(b)** .099   **(c)** .085   **(d)** .0095   **(e)** .15

6.1. **(a)** 1.523 **(b)** .127 **(c)** .046 **(d)** 1.077 **(e)** .333

7.1. 4.1

8.1. **(a)** 1.13 **(b)** .02 **(c)** .713 **(d)** 8.1 **(e)** 3

9.1. 230

9.2. **(a)** .710 **(b)** 3.333 **(c)** 3.333 **(d)** 2.857 **(e)** 852

10.1. **(a)** $6.24 **(b)** $76.47 **(c)** $3.01 **(d)** $72.35 **(e)** $7.01

13. WORD PROBLEMS

1. (2)
```
            15.5
$.68. ) $10.54.0
         6 8
         3 74
         3 40
           340
           340
```

2. (2)
```
           362.6
5.5 )1,994.3.0
      1 65
      344
      330
      143
      110
      330
      330
```

3. (1)
```
          11
.15. )1.65.
      1 5
        15
        15
```

4. (3)
```
          .5
10 )5.0
```

5. (5)
```
          $ 1.20
52 )$62.40
     52
     10 4
     10 4
```

6. (4)
```
             6.
$6.40. ) $38.40.
         38 40
```

7. (3)
```
         $.33
3 )$.99
```

8. (1)
```
             4.
$8.95 )$35.80.
       35 80
```

TWENTY-TWO

Percentages

1. WHAT IS A PERCENTAGE?

A percentage is a part of a whole, like a decimal or a fraction.
Here is the percent sign: %
It comes after the number: 50% means 50 percent.
How did the American people vote in the 1976 election?

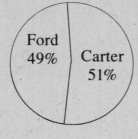

49% of the people voted for Ford,
51% voted for Carter.

1.1.

Any percentage can be changed to a decimal or to a fraction.
Examples:
49% = .49 = $^{49}/_{100}$ = 49 out of every 100
51% = .51 = $^{51}/_{100}$ = 51 out of every 100
50% = .50 = $^{50}/_{100}$ = 50 out of every 100

1.2.

To change a percentage to a decimal, here is what you do:
Example: change 75% to a decimal.
Drop the percent sign.

Starting from the right, count two places to the left and put a decimal point.

Your answer is .75.

75% ⟶ 75
.75

1.3.

You may have to put a zero to make up the two places.
Example: change 4% to a decimal.
Drop the percent sign.

Starting from the right, count two decimal places and put a decimal point.
You must add a zero to make up the second decimal place.

Your answer is .04.

1.4. Test Yourself

Change 5% to a decimal.
Check your answer on page 246. If it is right, go to section 1.5. If it is wrong or you don't know what to do, go to section 1.4A.

1.4A.

Follow the steps:
1. Drop the percent sign: 5
2. Starting from the right, count two places and put the decimal point: .05. You need to add a zero to make up the second decimal place.
If you found .5, it's because you forgot to add the zero.
You need it because you must have two places after the decimal point.

1.5. Practice

Change to decimals:

(a) 3%   (b) 75%   (c) 62%   (d) 7%   (e) 9%   (f) 48%   (g) 19%   (h) 36%

Check your answers on page 246.

1.6.

To change a percentage to a fraction, here is what you do:
Example: change 29% to a fraction.
Drop the percent sign.

Your number will be the numerator.

The denominator is always 100.
Your answer is $^{29}/_{100}$.

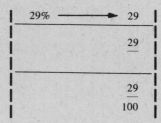

1.7. Practice

$75\% = .75 = {}^{75}/_{100} = 75$ our of every 100

(a) 23% =   =   =

(b) 15% =   =   =

(c) 16% =   =   =

(d) 8% =   =   =

(e) 30% =   =   =

Check your answers on page 246. If they are not all right, read sections 1 to 1.6 again.

2. FINDING A PERCENTAGE OF A WHOLE NUMBER

To solve word problems, you must change percentages to decimals.
Example: The sales tax on a $79 chair is 4%. What is the sales tax?
    4% of $79
The problem means that 4 cents of every dollar go for taxes.
The chair costs 79 dollars. How much of that amount goes for taxes?
Here is how to find out:
Change 4% to a decimal.
Multiply 79 × .04
Remember: there must be as many decimal places in the answer as
there are in the problem.

```
4% = .04

      79
   ×  .04
     316
     000
    3.16
```

Your answer is 3.16. The sales tax is $3.16.

2.1.

Here is another example: find 50% of 400.
Change 50% to a decimal.
Multiply 400 × .50

Your answer is 200. 50% of 400 is 200.

$$50\% = .50$$

$$\begin{array}{r} 400 \\ \times\ .50 \\ \hline 200.00 = 200 \end{array}$$

2.2. Practice

Find:
(a) 30% of 300   **(b)** 15% of 4,500   **(c)** 3% of 10   **(d)** 75% of 200
(e) 2% of 630

Check your answers on page 246.

3. PERCENTAGE REMINDERS

A percentage is a part of a hundred:
25% means 25 out of every 100
A percentage can be written as a decimal or as a fraction:
$25\% = .25 = {}^{25}/_{100}$
To change a percentage to a decimal:
drop the percent sign,
starting from the right, count two places and put a decimal point.
You may have to add a zero to make up the second decimal place.
$4\% = .04$
To change a percentage to a fraction:
drop the percent sign,
put 100 as a denominator.
$4\% = {}^{4}/_{100}$
To find a percentage of a number, change the percentage to a
decimal, and multiply the number by this decimal:
4% of $150 = 150 \times .04$

4. MORE PRACTICE

Write each as a decimal and as a fraction.
(a) 27% =   **(b)** 5% =   **(c)** 84% =   **(d)** 97% =   **(e)** 8% =

Check your answers on page 247.

4.1.

Find:
(a) 10% of 250   **(b)** 8% of $32   **(c)** 50% of $480   **(d)** 33% of $99
(e) 25% of $100

Check your answers on page 247.

5. WORD PROBLEMS—PERCENTAGES

Percentage word problems are easier to solve if you look for percentage word clues. Some percentage word clues are:

DISCOUNT—A $20 jacket is sold at a 10% discount. How much is the discount?

INTEREST—The bank pays you 6% interest on the $350 you save. How much interest does the bank pay you?

SALES TAX—The city sales tax is 8%. What is the sales tax on a $7.50 restaurant bill?

OF—60% of the 30 employees live outside the city and commute to work by car. How many employees is that?

DISCOUNT—Change 10% to a decimal.

$$
\begin{array}{r}
\$20 \\
\times\ .10 \\
\hline
00 \\
200 \\
\hline
\$2.00
\end{array}
$$

The discount is $2.00.

INTEREST—Change 6% to a decimal.

$$\begin{array}{r} \$350 \\ \times\ .06 \\ \hline 21\ 00 \\ 00\ 00 \\ \hline \$21.\ 00 \end{array}$$

The interest is $21.00.

SALES TAX—Change 8% to a decimal.

$$\begin{array}{r} \$7.50 \\ \times\ .\ 08 \\ \hline 60\ 0\ 0 \\ 00\ 0\ 0 \\ \hline \$.60\ 0\ 0 \end{array}$$

The sales tax is $ .60.

OF—Change 60% to a decimal.

$$\begin{array}{r} 30 \\ \times\ .60 \\ \hline 00 \\ 1800 \\ \hline 18.00 \end{array}$$

18 employees commute to work.

1. A mechanic bought a set of gauges for a gas torch for $150. The sales tax was 5%. How much sales tax did the mechanic pay?
   (1) $75   (2) $7.50   (3) $6.50   (4) $7.25
   (5) NT

2. 12 percent of the 360,700 employees of a nationwide company have been on the job more than 6 years. How many employees is that?
   (1) 43,284   (2) 44,384   (3) 4,384   (4) 4,328
   (5) NT

3. A savings bank pays 7% interest. What is the interest on $390?
   (1) $37.50   (2) $30   (3) $38.30   (4) 27.30
   (5) NT

4. A $9.00 sports shirt was marked down 30% in a summer clearance sale. What was the amount of the markdown?
   (1) $9.30   (2) $8.70   (3) $3.00   (4) $2.70
   (5) NT

5. 30% of 150 employees health insurance plans do not cover maternity expenses. How many plans is that?
   (1) 180   (2) 145   (3) 45   (4) 138
   (5) NT

6. American Electric pays disabled employees 60% of their weekly earnings. An employee who has a salary of $180 a week will receive how much in disability earnings?
   (1) $81   (2) $108   (3) $148   (4) 111
   (5) NT

7. Ms. Santiago pays $2,200 for qualified childcare expenses. 20% of that amount can be deducted for tax purposes. How much can Ms. Santiago take away from $2,200?
   (1) $440   (2) $42   (3) $222   (4) $2,200
   (5) NT

8. A quart of ice cream has 740 calories. Diet-Delight Ice Cream has 38% fewer calories. 38% of 740 calories is how many calories?
   (1) 280   (2) 210.5   (3) 240.9   (4) 281.2
   (5) NT

Check your answers on page 247. If all your problems are right, start Evaluation. If one or two are wrong, make sure you understand the right answers, and start Evaluation. If you made more than two mistakes, read Chapter 22 again. Then start Evaluation.

6. ANSWERS

1.4. .05

1.5. (a) .03   (b) .75   (c) .62   (d) .07   (e) .09

     (f) .48   (g) .19   (h) .36

1.7. (a) 23% = .23 = $^{23}/_{100}$ = 23 out of every 100

     (b) 15% = .15 = $^{15}/_{100}$ = 15 out of every 100

     (c) 16% = .16 = $^{16}/_{100}$ = 16 out of every 100

     (d) 8% = .08 = $^{8}/_{100}$ = 8 out of every 100

     (e) 30% = .30 = $^{30}/_{100}$ = 30 out of every 100

2.2. (a) 10   (b) 675   (c) .3   (d) 150   (e) 18.60

4. **(a)** $.27 = {}^{27}/_{100}$ **(b)** $.05 = {}^{5}/_{100}$ **(c)** $.84 = {}^{84}/_{100}$ **(d)** $.97 = {}^{97}/_{100}$

(e) $.08 = {}^{8}/_{100}$

4.1. **(a)** 25 **(b)** $2.56 **(c)** $240 **(d)** $32.67 **(e)** $25

5. **WORD PROBLEMS**

1. (2) $150
× .05
$7.50

2. (1) 360,700
× .12
721400
3607000
43,284.00

3. (4) $390
× .07
$27.30

4. (4) $9.00
× .30
$2.7000
$2.70

5. (3) 150
× .30
45.00

6. (2) $180
× .60
$108.00

7. (1) $2,200
× .20
$440.00

8. (4) 740
× .38
5920
22200
281.20

Evaluation: Addition, Subtraction, Multiplication, and Division of Decimals and Percentages

Add, Subtract, Multiply, or Divide. Work carefully. Watch the signs.

1. NUMBER PROBLEMS

(1) $\begin{array}{r} \$19.76 \\ -\ \ 8.43 \\ \hline \end{array}$

(2) $\begin{array}{r} .958 \\ .24 \\ +\ .19 \\ \hline \end{array}$

(3) $.5\overline{)87.65}$

(4) $\begin{array}{r} 15.95 \\ \times\ .042 \\ \hline \end{array}$

(5) $\begin{array}{r} .309 \\ \times\ .89 \\ \hline \end{array}$

(6) $\begin{array}{r} \$86.07 \\ -\ 39.48 \\ \hline \end{array}$

(7) $\begin{array}{r} \$138.90 \\ -\ 59.90 \\ \hline \end{array}$

(8) $.10\overline{)3.870}$

(9) $.52\overline{)26}$

(10) $\begin{array}{r} 3.47 \\ \times\ .06 \\ \hline \end{array}$

(11) $\begin{array}{r} 12.19 \\ -\ 8 \\ \hline \end{array}$

(12) $.65\overline{)3865}$

(13) $\begin{array}{r} 7.002 \\ \times\ 6.5 \\ \hline \end{array}$

(14) $\begin{array}{r} \$\ .88 \\ -\ \ .76 \\ \hline \end{array}$

(15) $\begin{array}{r} .59 \\ \times\ .16 \\ \hline \end{array}$

(16) $\begin{array}{r} 256.394 \\ 38.394 \\ +\ \ 9.08 \\ \hline \end{array}$

(17) $.009\overline{).7857}$

(18) $8.5\overline{)9.675}$

(19) $\begin{array}{r} 3.095 \\ +\ 7.104 \\ \hline \end{array}$

(20) $\begin{array}{r} .567 \\ +\ .819 \\ \hline \end{array}$

(21) $3.9\overline{).78}$

(22) $.34\overline{)170.68}$

(23) $\begin{array}{r} .648 \\ \times\ \ \ 5 \\ \hline \end{array}$

(24) $\begin{array}{r} 9.301 \\ +\ 2.54 \\ \hline \end{array}$

(25)   .895   (26)   7.1   (27)   7.42   (28)   21.42
           $\times$ 10 $-$ .9 $-$ 4.62 $\times$ .88

(29)   13.040
           $-$ 9.608

Round off each answer to the nearest dime.
(30) $26.57   (31) $189.76
           $+$ 48.29 $-$ 104.32

Write these percentages as decimals.
(32) 78%   (33) 19%   (34) 10%   (35) 3%   (36) 17%

Find:
(37) 15% of $720.00   (38) 80% of $800.00   (39) 5% of $680.00

WORD PROBLEMS

40. A pill bottle weighed 3.036 grams. Another weighed 2.489 grams. How many grams did both bottles weigh?
    (1) 23,525   (2) 5,489   (3) 5,525   (4) 6
    (5) NT

41. Electra-Deluxe floor cleaners increased $15.35 in price since last year. It was $169.95. What is the new price?
    (1) $200   (2) $195.95   (3) $185.95   (4) $148.95
    (5) NT

42. Marika used 2.9 yards of trim on the kitchen curtains, 3.7 yards on the living room curtains, and 1.5 yards on the bathroom curtain. How many yards of trim did she need in all?
    (1) 8.1   (2) 6.2   (3) 7.21   (4) 6.21
    (5) NT

43. You buy 8.5 gallons of gas on Monday. On Saturday you put in 12.2 gallons. How much more gas did you buy on Saturday than on Monday?
    (1) 4.8   (2) 4.7   (3) 3.8   (4) 3.7
    (5) NT

44. Jose went shopping with 100 dollars in his pocket. At the end of the day $37.26 remained. How much did Jose spend?
(1) $62.74   **(2)** $79.84   **(3)** $63.74   **(4)** $63.83
(5) NT

45. There were 8.75 pounds of cheese when the day cook went off duty. The next day there were 6.25 pounds of cheese. How many pounds of cheese had been used by the night cook?
(1) 3.15   **(2)** 2.5   **(3)** 2.55   **(4)** 1.5
(5) NT

46. A 10-lb. bag of charcoal briquets costs $1.49. 7 bags of charcoal would total how much?
(1) $10.66   **(2)** $9.56   **(3)** $9.28   **(4)** $10.43
(5) NT

47. Arnold received a $62.35 disability check each month. How much money does he receive in 12 months?
(1) $648.20   **(2)** $750.80   **(3)** $748.20   **(4)** $748.12
(5) NT

48. 8 people share 6 pounds of steak. What part of a pound is each person's share?
(1) .75   **(2)** .50   **(3)** .65   **(4)** 1.2
(5) NT

49. Amco Canning Company employs 900 people. 20% of the workers travel to the plant by car. How many people drive to work?
(1) 920   **(2)** 980   **(3)** 200   **(4)** 18
(5) NT

50. It costs $8.50 to drive on the turnpike from Ohio to New Jersey. This is 22% higher than the price last year. How much was the fare increased?
(1) $1.87   **(2)** $.22   **(3)** $8.72   **(4)** $9.00
(5) NT

Check your answers on page 251. If you made mistakes, make sure you understand the right answers. Go back to the chapters if you need to study them again.

ANSWERS

(1) $11.33	**(2)** 1.388	**(3)** 175.3	**(4)** 0.6699	**(5)** .27501
(6) $46.59	**(7)** $79.00	**(8)** 38.7	**(9)** 50	**(10)** .2082
(11) 4.19	**(12)** .595	**(13)** 45.513	**(14)** $ .12	**(15)** .0944
(16) 303.868	**(17)** 87.3	**(18)** 1.138	**(19)** 10.199	**(20)** 1.386
(21) .2	**(22)** 502	**(23)** 3.24	**(24)** 11.841	**(25)** 8.95
(26) 6.2	**(27)** 2.80	**(28)** 18.8496	**(29)** 3.432	**(30)** $74.90
(31) $85.40	**(32)** .78	**(33)** .19	**(34)** .10	**(35)** .03
(36) .17	**(37)** $180	**(38)** $640	**(39)** $34	

40. (3)
$$\begin{array}{r} 3.036 \\ + \ 2.489 \\ \hline 5.525 \end{array}$$

41. (5)
$$\begin{array}{r} \$169.95 \\ + \ \ \ 15.35 \\ \hline \$185.30 \end{array}$$

42. (1)
$$\begin{array}{r} 2.9 \\ 3.7 \\ + \ 1.5 \\ \hline 8.1 \end{array}$$

43. (4)
$$\begin{array}{r} 12.2 \\ - \ 8.5 \\ \hline 3.7 \end{array}$$

44. (1)
$$\begin{array}{r} \$100.00 \\ - \ \ \ 37.26 \\ \hline \$ \ 62.74 \end{array}$$

45. (2)
$$\begin{array}{r} 8.75 \\ - \ 6.25 \\ \hline 2.50 \end{array}$$

46. (4)
$$\begin{array}{r} \$1.49 \\ \times \ \ \ 7 \\ \hline \$10.43 \end{array}$$

47. (3) $62.35
            × 12
            12470
            62350
          $748.20

48. (1)

$$8\,\overline{)6.00}\;\; .75$$

          5 6
            40
            40

49.(5)   900
       × .20
      180.00

50. (1) $8.50
        × .22
       1700
      17000
     1.8700 = $ 1.87

TWENTY–THREE

Measurement: Length

1. WHAT IS A UNIT OF MEASUREMENT?

To measure things, we need units of measurement. Units of measurement are agreed upon so that everyone can know the amount that is being talked about.

In this chapter and in the following chapters, we'll talk about the American system, and about the metric system, which is already used all over the world. The United States will start using the metric system fully by the 1980's.

2. MEASURING LENGTH

Here are the American units to measure length. The short way to write them is inside the ( ).

Inch (in)   This is an inch: |_____|
Foot (ft)   12 inches make 1 foot
Yard (yd)   3 feet make 1 yard
Mile (mi)   1,760 yards make 1 mile

3. HOW TO USE LENGTH UNITS

Use inches to measure short things: pencils, books, silverware, nails, and pins.

Use feet to measure your height, the length of a wall, or a window.

Use yards to measure fabric for sewing, a football field, or a swimming pool.

Use miles to measure long distances: from the earth to the moon, from Trenton, N.J., to Atlanta, Ga.

4. CHANGING UNITS

One yard is 3 feet. But how many inches equal one yard?
You know that: 12 inches make 1 foot
So replace each foot by 12 inches:
Your answer: 1 yd = 36 in

How many feet are 48 inches?

You know that: 12 inches make 1 foot
So each time you have 12 inches, that makes 1 foot. How many times does 12 go into 48? 4 times. So 48 inches makes 4 feet.

When you change to a smaller unit, you multiply.
When you change to a larger unit, you divide.

1 yd = 3 ft
1 ft = 12 in
1 yd = 3 × 12 in
1 yd = 36 in

48 in = ? ft
12 in = 1 ft

$$12 \overline{)48}^{\,4}$$

48 in = 4 ft

4.1. Practice

Change the units.

(a) 96 in = _____ ft   (b) 9 yd = _____ in   (c) 12 ft = _____ yd

(d) 8,800 yd = _____ mi   (e) 9 ft = _____ in

Check your answers.

5. ADDING LENGTHS

To add lengths, start from the right: add inches, then feet, then yards.
Example: 3 yd 1 ft 4 in + 5 yd 1 ft 5 in

Put in column form with inches under inches, feet under feet, and yards under yards.
Add inches.

Add feet.

Add yards.
Your answer: 8 yards, 2 feet, 9 inches.

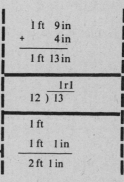

3 yd 1 ft 4 in
+ 5 yd 1 ft 5 in
_____
9 in

3 yd 1 ft 4 in
+ 5 yd 1 ft 5 in
_____
2 ft 9 in

3 yd 1 ft 4 in
+ 5 yd 1 ft 5 in
_____
8 yd 2 ft 9 in

5.1. If you get 12 inches or more in the inches column, change inches to feet.
Example: 1 ft 9 in + 4 in

Put in column form. Add inches and add feet.
Your answer: 1 ft 13 in.

But 13 inches is more than 12 inches. So change inches to feet. To do that, divide by 12. 13 inches make 1 foot and 1 inch.

Add the new foot to the foot you already have. 1 inch stays by itself.
Your final answer: 2 ft 1 in.

1 ft   9 in
+   4 in
_____
1 ft 13 in

        1 r 1
12) 13
_____
1 ft

1 ft   1 in
_____
2 ft 1 in

5.2. You may have to change inches to feet, or feet to yards.
Example: 3 yd 2 ft 7 in + 2 yd 2 ft 9 in

Put in column form.
Add inches, feet, and yards.
Your answer is 5 yd 4 ft 16 in.

But 16 inches is more than 12 inches.
Change inches to feet.
16 inches make 1 foot and 4 inches.
Your new answer is 5 yd 5 ft 4 in.

But 5 feet is more than 3 feet.
Change feet to yards.
5 feet make 1 yard and 2 feet.
Your final answer: 6 yd 2 ft 4 in.
Always reduce your answer as much as you can.

```
   3 yd  2 ft   7 in
 + 2 yd  2 ft   9 in
 ─────────────────────
   5 yd  4 ft  16 in
═════════════════════
   5 yd  4 ft
           1 ft   4 in
 ─────────────────────
   5 yd  5 ft   4 in
═════════════════════
   5 yd         4 in
   1 yd  2 ft
 ─────────────────────
   6 yd  2 ft   4 in
```

5.3. Practice

    (a)   8 in
         + 9 in

    (b)   2 yd 2 ft 9 in
         + 4 yd 2 ft 7 in

    (c) 3 yd 2 ft 10 in + 3 yd 2 ft 5 in =

Check your answers.

6. SUBTRACTING LENGTHS

To subtract lengths, start from the right: subtract inches, then feet, then yards.
Example: 5 yd 2 ft 10 in − 2 yd 1 ft 7 in

Put in column form.
Subtract inches, then feet, then yards.
Your answer: 3 yd 1 ft 3 in.
You can check it by adding.

```
  5 yd 2 ft 10 in
- 2 yd 1 ft  7 in
  ---------------
  3 yd 1 ft  3 in
```

6.1. You may have to borrow.
Example: 1 ft 3 in − 8 in
Put in column form.
Subtract inches . . . but 3 is less than 8.
Borrow 1 foot from the feet column.
Change it to 12 inches.
You now have 12 + 3 = 15 inches.
Subtract inches.
There is nothing to subtract in the feet column.
Your answer: 7 inches. You can check it by adding.
You may have to borrow more than once.

```
     1 ft  3 in
   -       8 in
   ---------

          15 in
   1̶f̶t̶ 3̶ i̶n̶
   -      8 in
   ---------
          7 in
```

6.2. Practice

(a)   5 ft 9 in
      − 2 ft 3 in
      _____

(b)   7 yd 2 ft 8 in
      − 3 yd 1 ft 5 in
      _____

(c) 8 ft 5 in − 5 ft 10 in =

Check your answers.

7. MULTIPLYING LENGTHS

Multiply units separately, then reduce the answer if necessary.
A short telephone cord is 2 ft 4 in. How long is a cord 4 times that length?

  2 ft 4 in × 4

Put in column form. Start from the right:
Multiply inches.

Multiply feet.
Your answer: 8 ft 16 in.

Change inches to feet. 16 in = 1 ft 4 in.
Your new answer: 9 ft 4 in.

Change feet to yards. 9 ft = 3 yd.
Your final answer: 3 yd 4 in.
Always reduce your answer as much as you can.

```
  2 ft  4 in
×        4
─────────────
        16 in

  2 ft  4 in
×        4
─────────────
  8 ft 16 in

  8 ft
  1 ft  4 in
─────────────
  9 ft  4 in

  3 yd   4 in
```

7.1. Practice

(a)   1 ft 3 in
    ×   3
    ─────────

(b)   2 ft 2 in
    ×   7
    ─────────

(c) 2 yd 2 ft 2 in × 4 =

Check your answers.

8. DIVIDING LENGTHS

A piece of fabric 6 feet 3 inches long is cut into 3 equal parts. How long is each piece?

6 ft 3 in ÷ 3

To divide lengths, use the other division sign.
Start from the left: divide feet.
Divide inches.
Your answer: 2 ft 1 in.
You can check it by multiplying.
Remember: when you divide, you always start from the *left*.

```
        2 ft
3 ) 6 ft 3 in

       2 ft 1 in
3 ) 6 ft 3 in
```

8.1. If you cannot start dividing, you must change feet to inches.
Example: 2 ft 3 in ÷ 3
Put the division in usual form.
Divide feet . . . but 3 doesn't go into 2.
So change feet to inches.
1 ft = 12 in, so 2 ft × 12 = 24 in.

Add these 24 inches and the 3 inches you already have. You now have 27 inches.
Divide inches.
Your answer: 9 in.
You can check it by multiplying.

```
3 ) 2 ft 3 in

2 ft = 24 in

24 in + 3 in = 27 in

       9 in
3 ) 27 in
```

8.2. If you get a remainder after you divide the first unit, change it to the next unit and continue dividing.
Example: 3 yd 1 ft ÷ 2
Put in usual form. Divide yards.
2 goes into 3, one time. Remainder: 1 yd.

Change the remainder to feet. 1 yd = 3 ft.
Add these feet to the 1 ft you already had.
You now have 4 ft.
Divide feet.
Your answer: 1 yd 2 ft.

```
        1 yd
   2 ) 3 yd  1 ft
      - 2
        1 yd
   ─────────────
        1 yd  2 ft
   2 ) 3 yd  1 ft
      - 2
        1 yd  3 ft
               4 ft
             - 4
               0
```

8.3. Practice

(a) 5) 2 ft 1 in   **(b)** 2) 1 ft 2 in   **(c)** 6 yd 2 ft ÷ 10 =

(d) 2 yd 2 ft ÷ 4 =

Check your answers.

9. MEASURING LENGTH THE METRIC WAY

Here are the principal metric units to measure length. The short way to write them is inside the ( ).

Centimeter (cm)   · This is a centimeter: └───┘
Meter (m)   100 centimeters make 1 meter
Kilometer (km)   1,000 meters make 1 kilometer.

10. HOW TO USE METRIC UNITS

Use centimeters to measure short things: paper clips, pens, nails, needles, glasses, or silverware.

Use meters to measure longer things: height of buildings, walls, swimming pools, football fields, or pieces of fabric.

Use kilometers to measure long distances: from New York to Boston, from Paris to Calcutta, and so on.

11. CHANGING UNITS

How many centimeters are there in 14 meters?

You know that: 100 centimeters make 1 meter.

So replace each meter by 100 cm:

Your answer: 14 m = 1400 cm

How many kilometers are there in 35,000 meters?

You know that: 1000 meters make 1 kilometer.

So each time you have 1,000 meters, you have 1 kilometer.

How many times does 1,000 go into 35,000?

35 times. So 35,000 m make 35 km.

$$1 \text{ m} = 100 \text{ cm}$$
$$14 \text{ m} = 14 \times 100 \text{ cm}$$
$$14 \text{ m} = 1400 \text{ cm}$$
$$1,000 \text{ m} = 1 \text{ km}$$
$$1,000 \overline{)35,000}^{\;35}$$
$$35,000 \text{ m} = 35 \text{ km}$$

Remember: when you change to a smaller unit, you multiply (add zeros), and when you change to a larger unit, you divide (you take away zeros).

11.1. Practice

Change the units.

    (a) 900 cm = _____ m   **(b)** 12,000 m = _____ km

    (c) 4 m = _____ cm   **(d)** 5 km = _____ m

Check your answers.

12. LENGTH REMINDERS

American system	*Metric system*
inch	centimeter
foot = 12 inches	meter = 100 centimeters
yard = 3 feet, or 36 inches	kilometer = 1,000 meters
mile = 1,760 yards	

To change to a smaller unit, you multiply.
To change to a larger unit, you divide.

12.1. More Practice

Change the units

    (a) 120 in = _____ ft   **(b)** 99 ft = _____ yd   **(c)** 600 cm = _____ m

    (d) 18,000 m = _____ km   **(e)** 16 yd = _____ ft

A. Add. Reduce where necessary

    (a)   1 ft 2 in
         +   9 in

    (b)   6 yd 2 ft 10 in
            1 yd 2 ft 9 in
         + 2 yd 2 ft 5 in

B. Subtract. Reduce when necessary

    (a)   81 cm
         − 63 cm

    (b)   9 yd 11 in
         − 6 yd 5 in

    (c)   2 ft 1 in
         −   8 in

    (d)   14 yd 8 in
         −   11 in

    (e)   8 yd 1 ft 6 in
         − 3 yd 2 ft 8 in

C. Multiply

    (a)   9 in
         × 9

    (b)   2 ft 8 in
         ×   3

    (c)   12 yd 1 ft 9 in
         ×   1

    (d) 2 yd 7 in
         × 6

    (e) 2 ft 10 in
         ×   5

D. Divide

    (a) 12) 48 m   **(b)** 5) 45 km   **(c)** 9) 1 ft 6 in

    (d) 5) 10 in   **(e)** 12) 3 yd 2 ft

Check your answers.

12.2. Word Problems

Before starting this section, review the word clues for addition, subtraction, multiplication, and division.

1. A bed frame puts a box spring 1 foot 7 inches from the floor. A box spring and mattress are 11 inches high. How high is it from the floor to the top of the mattress?
(1) 1 yd   **(2)** 2 ft 1 in   **(3)** 1 ft 15 in **(4)** 2 ft 6 in
(5) none of these

1. 1 2 3 4 5

2. Rafael is 6 feet 3 inches tall. How many inches is that?
(1) 63 in   **(2)** 75 in   **(3)** 79 in   **(4)** 66 in
(5) none of these

2. 1 2 3 4 5

3. The Christmas tree in Rockefeller Center in New York is 66 feet tall. How many yards is that?
(1) 21 yd   **(2)** 22 yd   **(3)** 23½ yd   **(4)** 21½ yd
(5) none of these

3. 1 2 3 4 5

4. A 1 foot 3 inch candle burned 8 inches. How high was it after that?
(1) 7 in   **(2)** 11 in   **(3)** 9 in   **(4)** 8 in
(5) none of these

4. 1 2 3 4 5

5. There are 8 cartons of oranges stacked in a grocery store. Each carton is 2 feet 3 inches high. How tall is the whole stack of cartons?
(1) 22 ft   **(2)** 6 yd   **(3)** 6 yd 2 ft **(4)** 6 yd 1 ft
(5) none of these

5. 1 2 3 4 5

Check your answers. If all your answers are right, start Chapter 18. If one or two are wrong, make sure that you understand the right answers and then start Chapter 18. If you made more than two mistakes, you should read Chapter 17 again before starting Chapter 18.

13. ANSWERS

4.1. **(a)** 8 ft   **(b)** 324 in   **(c)** 4 yd   **(d)** 5 mi   **(e)** 108 in

5.3. **(a)** 1 ft 5 in   **(b)** 7 yd 2 ft 4 in   **(c)** 7 yd 2 ft 3 in

6.2. **(a)** 3 ft 6 in = 1 yd 6 in   **(b)** 4 yd 1 ft 3 in   **(c)** 2 ft 7 in

7.1. **(a)** 3 ft 9 in = 1 yd 9 in   **(b)** 14 ft 14 in = 5 yd 2 in

       (c) 8 yd 8 ft 8 in = 10 yd 2 ft 8 in

8.3. **(a)** 5 in   **(b)** 7 in   **(c)** 2 ft   **(d)** 2 ft

11.1. **(a)** 9 m   **(b)** 12 km   **(c)** 400 cm   **(d)** 5,000 m

12.1. **(a)** 10 ft   **(b)** 33 yd   **(c)** 6 m   **(d)** 18 km   **(e)** 48 ft

A. **(a)** 1 ft 11 in   **(b)** 11 yd 2 ft

B. **(a)** 18 cm   **(b)** 3 yd 6 in   **(c)** 1 ft 5 in   **(d)** 13 yd 2 ft 9 in

(e) 4 yd 1 ft 10 in

C. **(a)** 2 yd 9 in   **(b)** 2 yd 2 ft   **(c)** 12 yd 1 ft 9 in   **(d)** 13 yd 6 in   **(e)** 4 yd 2 ft 2 in

D. **(a)** 4 m   **(b)** 9 km   **(c)** 2 in   **(d)** 2 in   **(e)** 11 in

12.2. Word Problems

1. (4)   1 ft   7 in
       +   11 in
       ―――――――――――
          1 ft 18 in = 2 ft 6 in

2. (2) 6 ft 3 in   12 × 6 = 72
        72 + 3 = 75 in

3. (2) 66 ft   3 ft = 1 yd

            22 yds
        3) 66

4. (1) 1 ft 3 in = 15 in
                   − 8 in
             ―――――――――
                     7 in

5. (2)   2 ft   3 in
        ×   8 in
        ―――――――――――
        16 ft 24 in = 18 ft = 6 yd

TWENTY–FOUR

Measurement: Weight

1. MEASURING WEIGHT

Here are the American units to measure weight. The short way to write them is inside the ( ).

Ounce (oz)   a candy bar weighs about 1 ounce

Pound (lb)   16 ounces make 1 pound

2. HOW TO USE WEIGHT UNITS

Use ounces to weigh small amounts: spices, teas, candies, or tacks.

Use pounds to measure your weight, the weight of a car, bricks, furniture, or meat.

3. CHANGING UNITS

To change pounds to ounces, multiply the number of pounds by 16.

To change ounces to pounds, divide the number of ounces by 16.

Example: How many ounces are there in 5 pounds?

You know that: 16 ounces make 1 pound.

Replace each pound by 16 ounces:

Your answer: 5 lb = 80 oz

How many pounds are in 24 ounces?

You know that: 16 ounces make 1 pound.

Each time you have 16 ounces, you have 1 pound.

How many times does 16 go into 24? Once, with remainder 8.
Your answer: 24 oz = 1 lb 8 oz

$$1\,lb = 16\,oz$$
$$5\,lb = 5 \times 16\,oz$$
$$5\,lb = 80\,oz$$

$$16\,oz = 1\,lb$$
$$16\overline{)24}\;^{1r8}$$

3.1. Practice

Change to ounces:

    (a) 8 lb   **(b)** 7 lb   **(c)** 2 lb   **(d)** 3 lb   **(e)** 4 lb

Change to pounds:

    (f) 22 oz   **(g)** 32 oz   **(h)** 49 oz   **(i)** 19 oz   **(j)** 45 oz

Check your answers.

4. ADDING WEIGHTS

To add weights, start from the right: add ounces, then pounds.
One box of soap powder weighs 4 lb 9 oz, and another box weighs 9 lb 9 oz. How much
is that altogether?
Use column form.
Add ounces, then add pounds.
Your answer: 13 lb 18 oz.
18 ounces is more than 16 ounces. So change 18 oz to pounds:
18 oz = 1 lb 2 oz

Your final answer: 14 lb 2 oz.
Always reduce your answer if you can.

$$
\begin{array}{r}
4\,lb \quad 9\,oz \\
+\;9\,lb \quad 9\,oz \\
\hline
13\,lb \quad 18\,oz \\
\hline
13\,lb \\
1\,lb \quad 2\,oz \\
\hline
14\,lb \quad 2\,oz
\end{array}
$$

4.1. Practice

Add and reduce:

(a)	12 oz	**(b)**	3 lb 4 oz	**(c)**	8 lb 15 oz	**(d)** 9 lb 10 oz + 2 lb 9 oz =
	+ 13 oz		+ 2 lb 13 oz		+ 7 lb 8 oz	**(e)** 6 lb 14 oz + 9 lb 12 oz =

Check your answers.

5. SUBTRACTING WEIGHTS

To subtract weights, start from the right: subtract ounces, then pounds.
Example: 8 lb 10 oz – 6 lb 4 oz
Use column form.
Subtract ounces, then subtract pounds.
Your answer: 2 lb 6 oz.
You can check it by adding.

```
  8 lb 10 oz
– 6 lb  4 oz
  ─────────
  2 lb  6 oz
```

5.1. You may have to borrow.
Example: 9 lb 4 oz – 3 lb 12 oz
Put in column form.
Subtract ounces . . . but 4 is less than 12.

So borrow 1 pound from the pounds column.
Change it to 16 ounces.
You now have 16 + 4 = 20 ounces.
Subtract ounces, then subtract pounds.
Your answer: 6 lb 8 oz.

```
  9 lb  4 oz
– 3 lb 12 oz
  ─────────

  8    20
  9̶ l̶b̶  4̶ o̶z̶
– 3 lb 12 oz
  ─────────
  6 lb  8 oz
```

5.2. Practice

(a)   15 oz   **(b)**   4 lb   12 oz   **(c)**   57 lb   5 oz   **(d)** 27 lb 2 oz – 14 lb 14 oz =
     –   8 oz   – 1 lb   3 oz   – 49 lb 13 oz

Check your answers.

6. MULTIPLYING WEIGHTS

Multiply units separately, then reduce the answer if necessary.
Example: 5 lb   8 oz × 3
Use column form.
Multiply ounces, then multiply pounds.
Your answer: 15 lb   24 oz.
24 ounces is more than 16 ounces. So change ounces to pounds.
24 oz = 1 lb   8 oz.
Your final answer: 16 lb   8 oz.

```
        5 lb   8 oz
     ×          3
     ─────────────────
       15 lb  24 oz
     ═════════════════
       15 lb
        1 lb   8 oz
     ─────────────────
       16 lb   8 oz
```

6.1. Practice

(a) 7 lbs   2 oz × 5 =   **(b)** 1 lb   5 oz × 3 =   **(c)** 10 lbs   10 oz × 10 =

(d) 57 lbs   5 oz × 8 =   **(e)** 21 lbs   3 oz × 3 =

Check your answers.

7. DIVIDING WEIGHTS

Start from the left: divide pounds, then divide ounces.
Example: 8 lb   12 oz ÷ 4
Put in usual form.

Divide pounds, then divide ounces.

Your answer: 2 lb 3 oz.

You can check it by multiplying.

```
        2 lb  3 oz
   4 ) 8 lb 12 oz
```

7.1. If you cannot start dividing, change pounds to ounces.

4 people share 3 lb 4 oz of steak. What is each person's share?

    3 lb 4 oz ÷ 4

Start from the left.

Put in the usual division form. Divide pounds . . . but 4 doesn't go into 3.

So change pounds to ounces.

Add these 48 ounces to the 4 you already had.

You now have 52 ounces.

Divide ounces. Your answer: 13. Each person gets 13 ounces of steak.

```
   4 ) 3 lb 4 oz

   3 lb = 48 oz

   48 + 4 = 52

        13 oz
   4 ) 52 oz
```

7.2. If you get a remainder after you divide the first unit, change it to the next unit and continue dividing.

Example: 7 lb 2 oz ÷ 3

Put in the usual form.

Divide pounds. 3 goes into 7 two times.

Remainder: 1.

Change the remainder to ounces. Add to what you already had.

You now have 18 ounces.

Divide ounces.

Your answer: 2 lb 6 oz.

```
         2 lb   6 oz
   3 ) 7 lb   2 oz
     - 6
       1 lb  16 oz

            18 oz
          - 18
             0
```

7.3. Practice

(a) $5 \overline{\smash{)}\ 10\,\text{lb}\ 5\,\text{oz}}$ (b) $7 \overline{\smash{)}\ 5\,\text{lb}\ 4\,\text{oz}}$ (c) 3 lb 8 oz ÷ 4 =

(d) 8 lb 12 oz ÷ 7 =

Check your answers. If they are not all right, read section 7 again very carefully.

8. MEASURING WEIGHT THE METRIC WAY

Here are the principal metric units to measure weight. The short way to write them is inside the ( ).

Gram (g)   a pin weighs about 1 gram
Kilogram (kg)   1,000 grams make 1 kilogram

9. HOW TO USE METRIC UNITS

Use grams for very light objects: a candy weighs about 10 grams; a can of soup weighs 315 grams.

Use kilograms for heavier things: a newborn baby weighs about 4 kilograms; a book can weigh about 2 kilograms.

10. CHANGING UNITS

To change grams to kilograms, divide by 1,000 (that is, take away 3 zeros).
To change kilograms to grams, multiply by 1,000 (that is, add three zeros).
Examples: 3 kg = 3,000 g
            28,000 g = 28 kg

10.1. Practice

(a) 75,000 g = _____ kg (b) 21,000 g = _____ kg (c) 17,000 g = _____ kg

(d) 7 kg = _____ g   (e) 11 kg = _____ g

Check your answers.

11. WEIGHT REMINDERS

American system

ounce
pound = 16 ounces

Metric system

gram
kilogram = 1,000 grams

To change kilograms to grams, multiply by 1,000.
To change grams to kilograms, divide by 1,000.
To change pounds to ounces, multiply by 16.
To change ounces to pounds, divide by 16.

11.1. More Practice

(a) 35 lb 13 oz
   − 12 lb 15 oz

(b) 14 lb 9 oz
   ×   6

(c)   7 lb 5 oz
   13 lb 12 oz
   10 lb 9 oz
   + 4 lb 14 oz

(d) 9) 12 lb 6 oz

(e) 12 lb 10 oz
   ×   8

(f)   2 lb 15 oz
   ×   3

(g)   14 oz
   × 8

(h) 19 lb 4 oz
   − 8 lb 11 oz

(i)   15 lb
   − 8 oz

(j) 14) 16 lb 10 oz

Change the units:

(a) 4 lb 12 oz = _____ oz

(b) 38 oz = _____ lb _____ oz

(c) 6,000 g = _____ kg

(d) 4 kg = _____ g

11.2. Word Problems

Before starting this section, review the word clues for addition, subtraction, multiplication, and division.

1. A deep sea fisherman caught an 18 lb 9 oz codfish and a 93 lb 8 oz white marlin. How much did both fish weigh together?

1. 1 2 3 4 5

(1) 112 lb 1 oz   (2) 108 lb 1 oz
(3) 111 lb 10 oz   (4) 148 lb   (5) none of these

2. Ready Made Chow Mein comes in a package that weighs 2 lb 10 oz. It comes in 5 varieties. What is the total weight if you buy one of each kind?
   (1) 15 lb   **(2)** 10 lb 45 oz   **(3)** 13 lb 2 oz
   (4) 13 lb   **(5)** none of these

2.   1   2   3   4   5

3. 5 candy bars weighed 1 lb 4 oz altogether. Each had the same weight. What was the weight of each?
   (1) 8 oz   **(2)** 2 oz   **(3)** 4 oz   **(4)** 5 oz
   (5) none of these

3.   1   2   3   4   5

4. How many ounces are there in 46 pounds?
   (1) 636   **(2)** 736   **(3)** 616   **(4)** 763
   (5) none of these

4.   1   2   3   4   5

5. Esther had a watermelon that weighed 8 lb 5 oz. She wanted 7 people to have exactly the same amount each. What was the weight of each person's share?
   (1) 1 lb 3 oz   **(2)** 1 lb 5 oz
   (3) 1 lb 7 oz   **(4)** 13 oz   **(5)** none of these

5.   1   2   3   4   5

Check your answers. If all your answers are right, start Chapter 25. If one or two are wrong, make sure that you understand the right answers, and then start Chapter 25. If you made more than two mistakes, you should read Chapter 24 again before starting Chapter 25.

12. ANSWERS

3.1. **(a)** 128 oz   **(b)** 112 oz   **(c)** 32 oz   **(d)** 48 oz   **(e)** 64 oz

     (f) 1 lb 6 oz   **(g)** 2 lb   **(h)** 3 lb 1 oz   **(i)** 1 lb 3 oz   **(j)** 2 lb 13 oz

4.1. **(a)** 25 oz = 1 lb 9 oz   **(b)** 5 lb 17 oz = 6 lb 1 oz   **(c)** 15 lb 23 oz = 16 lb 7 oz

     (d) 11 lb 19 oz = 12 lb 3 oz   **(e)** 1 lb 26 oz = 16 lb 10 oz

5.2. **(a)** 7 oz   **(b)** 3 lb 9 oz   **(c)** 7 lb 8 oz   **(d)** 12 lb 4 oz

6.1. **(a)** 35 lb 10 oz   **(b)** 3 lb 15 oz   **(c)** 106 lb 4 oz   **(d)** 458 lb 8 oz

     (e) 63 lb 9 oz

7.3. **(a)** 2 lb 1 oz   **(b)** 12 oz   **(c)** 14 oz   **(d)** 1 lb 4 oz

10.1. **(a)** 75 kg   **(b)** 21 kg   **(c)** 17 kg   **(d)** 7,000 g   **(e)** 11,000 g

11.1. **(a)** 22 lb 14 oz   **(b)** 84 lb 54 oz = 87 lb 6 oz

(c) 34 lb 40 oz = 36 lb 8 oz **(d)** 1 lb 6 oz   **(e)** 96 lb 80 oz = 101 lb

(f) 6 lb 45 oz = 8 lb 13 oz   **(g)** 112 oz = 7 lb   **(h)** 10 lb 9 oz

(i) 14 lb 8 oz   **(j)** 1 lb 3 oz

Change the units:

(a) 76 oz   **(b)** 2 lb 6 oz   **(c)** 6 kg   **(d)** 4,000 g

11.2. Word Problems

1. (1)   18 lb   9 oz
      + 93 lb   8 oz
      ────────────
      111 lb 17 oz =
      112 lb   1 oz

2. (3)   2 lb 10 oz
          ×   5
      ────────────
      10 lb 50 oz =
      13 lb   2 oz

3. (3) 5) 1 lb 4 oz =

              4 oz
          5) 20 oz

4. (2)   46
          × 16
      ────────
          276
          460
      ────────
          736

5. (1)   1 lb 3 oz
          7) 8 lbs 5 oz
                 7
              ──────
              1 lbs 5 oz = 21 oz
                        21
                      ──────

TWENTY–FIVE

Measurement: Liquid

1. MEASURING LIQUIDS

Here are the American units to measure liquids. The short way to write them is inside the ( ).

Liquid ounce (oz)	Vanilla extract comes in a 2 ounce bottle
Cup (c)	8 liquid ounces make 1 cup
Pint (pt)	2 cups make 1 pint
Quart (qt)	2 pints make 1 quart
Gallon (gal)	4 quarts make 1 gallon

2. HOW TO USE LIQUID UNITS

You know these things, which you see everyday:

3. CHANGING UNITS

When you change to a smaller unit, you multiply.
When you change to a larger unit, you divide.
Example: how many cups are in 4 pints?
You know that: 2 cups make 1 pint.
Replace each pint by 2 cups:
Your answer: 8 c
How many gallons are in 18 quarts?

You know that: 4 quarts make 1 gallon.

Each time you have 4 quarts, you have 1 gallon.

How many times does 4 go into 18? 4 times, remainder 2.

So 18 quarts make 4 gallons, 2 quarts.

$$1 \, pt = 2 \, c$$
$$4 \, pt = 4 \times 2 \, c$$
$$4 \, pt = 8 \, c$$
$$18 \, qt = ? \, gal$$
$$4 \, qt = 1 \, gal$$
$$\begin{array}{r} 4r2 \\ 4 \overline{) \, 18} \end{array}$$
$$18 \, qt = 4 \, gal \, 2 \, qt$$

3.1. Practice

Change to cups:

    (a) 6 pt   **(b)** 10 pt   **(c)** 8 pt   **(d)** 3 pt   **(e)** 5 pt

Change to pints:

    (f) 12 c   **(g)** 6 c   **(h)** 8 c   **(i)** 3 c

Change to pints:

    (j) 5 qt   **(k)** 8 qt   **(l)** 3 qt   **(m)** 2 qt

Change to quarts:

    (n) 12 pt   **(o)** 18 pt   **(p)** 8 pt   **(q)** 15 pt

Change to gallons:

    (r) 8 qt   **(s)** 24 qt   **(t)** 12 qt   **(u)** 20 qt

Change to quarts:

    (v) 4 gal   **(w)** 7 gal   **(x)** 2 gal   **(y)** 6 gal

Check your answers.

4. ADDING LIQUID MEASURES

To add liquid measures, start from the right and add each unit separately. Always reduce your answer when you can.

Example: 3 gal 2 qt 1 pt + 4 gal 3 qt 1 pt
Use column form.
Add pints, then quarts, then gallons.
Your answer: 7 gal 5 qt 2 pt

Change pints to quarts: 2 pt = 1 qt
Your new answer: 7 gal 6 qt

Change quarts to gallons: 6 qt = 1 gal 2 qt
Your final answer: 8 gal 2 qt

```
  3 gal 2 qt 1 pt
+ 4 gal 3 qt 1 pt
──────────────────
  7 gal 5 qt 2 pt
──────────────────
  7 gal 5 qt
        1 qt
──────────────────
  7 gal 6 qt
──────────────────
  7 gal
  1 gal 2 qt
──────────────────
  8 gal 2 qt
```

4.1. Practice

Add and reduce:

    (a)   6 gal 2 qt 1 pt   **(b)**   3 qt 1 pt   **(c)** 7 gal 3 qt + 2 gal 3 qt =
          + 2 gal 3 qt 1 pt   + 3 qt
          ────────────────   ──────────

Check your answers.

5. SUBTRACTING LIQUID MEASURES

Subtract each unit separately. You may have to borrow.

Example: 7 gal 2 qt – 3 gal 3 qt
Use column form. Subtract quarts . . . but 2 is less than 3.
Borrow 1 gallon from the gallons column. Change it to 4 quarts.
You now have 2 + 4 = 6 quarts.
Subtract quarts, then subtract gallons.

Your answer: 3 gal 3 qt.

```
  7 gal 2 qt
- 3 gal 3 qt
────────────
    6   6
  7̶ ̶g̶a̶l̶ ̶2̶ ̶q̶t̶
- 3 gal 3 qt
────────────
  3 gal 3 qt
```

5.1. Practice

(a)   3 gal 1 qt
        – 1 gal 3 qt

(b)   3 qt
        –   1 pt

(c)   6 gal 2 qt 1 pt
        – 3 gal 3 qt 2 pt

Check your answer.

6. MULTIPLYING LIQUID MEASURES

A recipe says: "Add 1 pint 1 cup of milk." If you double the recipe, how much milk do you add?

1 pt 1 c × 2

Use column form.

Multiply cups, then multiply pints.

Your answer: 2 pt 2 c

Change cups to pints. Your new answer: 3 pt.

Change pints to gallons.   Your final answer: 1 qt 1 pt.

```
  1 pt 1 c
×       2
──────────
  2 pt 2 c
──────────
    2 pt
    1 pt
──────────
    3 pt
──────────
    3 pt
  1 qt 1 pt
```

6.1. Practice

Multiply and reduce:

(a)   3 gal 2 qt
        ×   5

(b) 4 gal 2 qt 1 pt
        ×   6

(c) 4 gal 2 qt
        ×   4

Check your answers.

7. DIVIDING LIQUID MEASURES

Start from the left: divide gallons, then quarts, then pints, and so on.

Example: 5 gal 1 qt ÷ 3

Use the other division sign.

3 goes into 5 once, remainder 2.

Change the remainder to quarts. You now have 9 qt.

Divide quarts.

Your answer: 1 gal 3 qt

```
            1 gal
      3 ) 5 gal 1 qt
          - 3
          ─────
            2 gal
```

```
            1 gal 3 qt
      3 ) 5 gal 1 qt
          - 3
          ─────
            2 gal 8 qt
                9 qt
              - 9
              ─────
                0
```

7.1. If you cannot start dividing, change to the next unit.

Example: 3 gal ÷ 4

Put in usual form. Divide gallons . . . but 4 doesn't go into 3.

Change gallons to quarts. You now have 12 qt.

Divide quarts. Your answer: 3 qt.

```
      4 ) 3 gal
```

```
            3 qt
      4 ) 12 qt
```

7.2. Practice

   (a) 2) 8 gal 2 qt   **(b)** 6) 4 gal 2 qt   **(c)** 5) 2 qt 1 pt

Check your answers.

8. LIQUID REMINDERS

American system   *Metric system*

Liquid ounce   Liter

Cup = 8 ounces

Pint = 2 cups

Quart = 2 pints

Gallon = 4 quarts

To change to a smaller unit, multiply.
To change to a larger unit, divide.

8.1. More Practice

Change the units

(a) 7 gal = _____ qt   **(b)** 20 pt = _____ qt   **(c)** 4 c = _____ qt

(d) 3 qt = _____ pt   **(e)** 17 qt = _____ gal _____ qt

Add, subtract, multiply, or divide. Reduce when necessary.

(f)   3 qt
        × 12
    _____

(g)   51 gal
           71 gal
        + 31 gal
        _____

(h)   2 gal 3 qt
        −   3 qt 1 pt
        _____

(i)   1 pt 1 c
         3 qt 1 pt 1 c
       + 2 qt 1 pt 1 c
       _____

(j) 8) 4 gal

Check your answers.

8.2. Word Problems

1. 6 cups will fill how many pints, quarts or gallons?
   (1) 2 pints   **(2)** 3 quarts   **(3)** 1 quart and 1 pint
   (4) 1 gallon   **(5)** none of these

   1. 1 2 3 4 5

2. A family drinks 1 gal 1 qt of milk a day. How much milk do they use in 5 days?
   (1) 10 qts   **(2)** 20 qts   **(3)** 2 gal 3 qt **(4)** 6 gal 1 qt
   (5) none of these

   2. 1 2 3 4 5

3. 1 cup of frozen lemonade goes into a pitcher. How many cups of water do you add to make 2 quarts of lemonade?
   (1) 7 c   **(2)** 10 c   **(3)** 4 c   **(4)** 5 c
   (5) none of these

   3. 1 2 3 4 5

4. Add 1 quart of rum, 2 gallons 2 quarts of orange juice and 1 quart of tonic water for summer punch. How much punch does this recipe make?
   (1) 4 gal   **(2)** 3 gal   **(3)** 2 gal 3 qt
   (4) 3 gal 2 qt   **(5)** none of these

   4. 1 2 3 4 5

5. Alice gives her violets 1 cup of water, her rubber tree 3 quarts 1 pint of water, her spider plant 1 pint 1 cup of water. How much water in all does she give her plants?

    5. 1 2 3 4 5

(1) 1 gal 1 pt   **(2)** 1 gal 2 qt   **(3)** 9 c

(4) 4 qt   **(5)** none of these

Check your answers. If all your answers are right, start Chapter 26. If one or two are wrong, make sure that you understand the right answers before starting Chapter 26. If you made more than two mistakes, read Chapter 25 again and then start Chapter 26.

9. ANSWERS

3.1. **(a)** 12 c   **(b)** 20 c   **(c)** 16 c   **(d)** 6 c   **(e)** 10 c   **(f)** 6 pt

      (g) 3 pt   **(h)** 4 pt   **(i)** 1½ pt   **(j)** 10 pt   **(k)** 16 pt   **(l)** 6 pt

      (m) 4 pt   **(n)** 6 qt   **(o)** 9 qt   **(p)** 4 qt   **(q)** 7½ qt   **(r)** 2 gal

      (s) 6 gal   **(t)** 3 gal   **(u)** 5 gal   **(v)** 16 qt   **(w)** 28 qt   **(x)** 8 qt

      (y) 24 qt

4.1. **(a)** 9 gal 2 qt   **(b)** 1 gal 2 qt 1 pt   **(c)** 10 gal 2 qt

5.1. **(a)** 1 gal 2 qt   **(b)** 2 qt 1 pt   **(c)** 2 gal 2 qt 1 pt

6.1. **(a)** 15 gal 10 qt = 17 gal 2 qt   **(b)** 24 gal 12 qt 6 pt = 27 gal 3 qt

      (c) 16 gal 8 qt = 18 gal

7.2. **(a)** 4 gal 1 qt   **(b)** 3 qt   **(c)** 1 pt

8.1. **(a)** 28 qt   **(b)** 10 qt   **(c)** 1 qt   **(d)** 6 pt   **(e)** 4 gal 1 qt

      (f) 9 gal   **(g)** 153 gal   **(h)** 1 gal 3 qt 1 pt   **(i)** 1 gal 3 qt 1 c

      (j) 2 qt

8.2. Word Problems

    1. (3) 6 c = 3 pt
            3 pt = 1 qt 1 pt
            6 c = 1 qt 1 pt

    2. (4)   1 gal 1 qt
         ×   5
         —————————
         5 gal 5 qt = 6 gal 1 qt

3. (1) 4 c = 1 qt   8 c

      8 c = 2 qt   − 1 c lemonade

      1 c lemonade   7 c water

4. (2)   2 gal 2 qt

            1 qt

    +   1 qt

      2 gal 4 qt = 3 gal

5. (1)   1 c

      3 qt 1 pt

    +   1 pt 1 c

      3 qt 2 pt 2 c = 1 gal 1 pt

TWENTY-SIX

Measurement: Time

1. MEASURING TIME

The units to measure time are:

Second (s)	It takes about one second to say "OH!"
Minute (min)	60 seconds make 1 minute
Hour (h)	60 minutes make 1 hour
Day	24 hours make a day
Week	7 days make a week

52 weeks plus one day make one year. Each year has 365 days, except every 4 years, when it has 366 days.

2. CHANGING UNITS

When you change to a smaller unit, you multiply.
When you change to a larger unit, you divide.
Example: how many minutes are in 5 hours?
You know that: 60 minutes make 1 hour.
Replace each hour by 60 minutes:
your answer: 5 hours is 300 minutes
How many weeks are in 84 days?
Each time you have 7 days, you have a week.
How many times does 7 go into 84? 12 times.
84 days make 12 weeks.

$$1\,h = 60\,min$$
$$5\,h = 5 \times 60\,min = 300\,min$$
$$84\,days = ?\,weeks$$
$$7\,days = 1\,week$$
$$7\overline{)84}^{\,12}$$

2.1. Practice

Change the units.

    (a) 180 s = _____ min   (b) 13 min = _____ s   (c) 300 min = _____ h

    (d) 120 h = _____ days   (e) 49 days = _____ wks

Check your answers.

3. ADDING TIME

Start from the right and add each unit separately. Always reduce your answer when you can.

Example: A car race lasted 5 hours 45 minutes and 20 seconds.
                Another race lasted 6 hours 35 minutes and 10 seconds.
                How long did the two races last altogether?
                  5 h 45 min 20 s + 6 h 35 min 10 s

Use column form.

Add seconds, then minutes, then hours.

Your answer: 11 h 80 min 30 s.

80 minutes is more than 60 minutes. Change minutes to hours.

80 min = 1 h 20 min

Your final answer: 12 h 20 min 30 s

```
    5 h 45 min 20 s
 +  6 h 35 min 10 s
   ────────────────
   11 h 80 min 30 s

   11 h         30 s
    1 h 20 min
   ────────────────
   12 h 20 min 30 s
```

3.1. Practice

Add and reduce:

    (a)   55 min   (b)   2 days 18 h   (c) 2 weeks 2 days + 6 days =
        + 49 min   + 3 days 12 h   (d) 1 h 20 min + 3 h 40 min =

Check your answers.

4. SUBTRACTING TIME

Subtract each unit separately. You may have to borrow.

A movie is 2 hours 15 minutes long. Another is 1 hour 40 minutes long.

How much longer is the first movie?

        2 h 15 min − 1 h 40 min

Put in column form.

Subtract minutes . . . but 15 is less than 40.

Borrow 1 hour from the hours column.

Change it to 60 minutes. You now have 60 + 15 = 75 minutes.

Subtract minutes, then subtract hours.

Your answer: the first movie is 35 minutes longer.

```
        2 h 15 min
      − 1 h 40 min

        1    75
        2 h  15 min
      − 1 h 40 min
        0 h 35 min = 35 min
```

4.1. Practice

(a)   5 min 45 s
      − 2 min 35 s

(b)   15 min 37 s
      − 12 min 48 s

(c)   3 h 15 min 30 s
      − 1 h 50 min 45 s

(d) 6 years 2 weeks − 2 years 1 week =

Check your answers.

5. MULTIPLYING TIME

Multiply each unit separately. Reduce when necessary.

A building takes 3 weeks 2 days and 4 hours to be finished. How long does it take to finish 4 buildings?

        3 weeks 2 days 4 h × 4

Use column form.

Multiply hours, then days, then weeks.

Your answer: 12 weeks 8 days 16 hours.

8 days is more than 7 days. So change days to weeks.

8 days = 1 week 1 day.

Your final answer: 13 weeks 1 day 16 hours.

```
        3 weeks  2 days   4 h
  ×                        4
  ────────────────────────────
       12 weeks  8 days  16 h

       12 weeks         16 h
        1 week  1 day
  ────────────────────────────
       13 weeks  1 day  16 h
```

5.1. Practice

(a)   45 min   (b) 2 weeks 2 days   (c) 2 weeks 5 days × 2 =
      × 3   ×   2   (d) 6 h 4 min 12 s × 4 =
      ───────   ──────────────

Check your answers.

6. DIVIDING TIME

Start from the left. Divide each unit separately. If you have a remainder, change it to
the next unit. If you cannot start dividing, change to next unit.

Example: 18 h 24 min 24 s ÷ 8

Put in usual form. Divide hours. Remainder: 2 h.

Change the remainder to minutes. 2 h = 120 min

You now have 144 min.

Divide minutes, then divide seconds.

Your answer: 2 h 18 min 3 s

```
            2 h
  8 ) 18 h 24 min 24 s
     - 16
      ────
       2 h


         2 h 18 min  3 s
  8 ) 18 h 24 min 24 s
     - 16
      ────
       2 h = 120 min
      ───────────────
           144 min
```

6.1. Practice

(a) 8) 16 h 48 s   (b) 12) 48 h 24 min   (c) 27 weeks 9 days ÷ 9 =
                                                     (d) 3 weeks 18 h ÷ 3 =

Check your answers.

7. TIME REMINDERS

Second
Minute = 60 seconds
Hour = 60 minutes
Day = 24 hours
Week = 7 days
To change to a smaller unit, multiply.
To change to a larger unit, divide.

7.1. More Practice

(a) 68 min = _____ h _____ min   (b) 540 s = _____ min

(c) 144 h = _____ days   (d) 56 weeks = _____ years _____ weeks

(e) 8 min = _____ s

Add, subtract, multiply, and divide. Reduce when necessary.

(f)   7 weeks 5 days   (g)   3 mn 28 s   (h)   1 year 6 days
   + 12 weeks 6 days   −   49 s   ×   7
   _____   _____   _____

(i) 9) 18 h 45 min 54 s   (j)   5 days 1 h
                                −   12 h
                                _____

Check your answers.

7.2. Word Problems

1. The first act of a play lasted 1 hour 15 minutes. There was a 20 minute intermission. The second act lasted 50 minutes. How long was the play including intermission?
   (1) 2 h 85 min   (2) 2 h 25 min   (3) 1 h 70 min
   (4) 2 h 30 mn   (5) none of these

   1. 1 2 3 4 5

2. Manny rode a subway for 40 minutes, then walked 25 minutes more to get to work. How much more time did he spend on the subway?
   (1) 1 hr 5 min   (2) 15 min   (3) 25 min   (4) 35 min
   (5) none of these

   2. 1 2 3 4 5

3. Pete had a short part-time job. He worked 3 hours a day for a week. How many hours in all did Pete work?
   (1) 21 h   **(2)** 24 h   **(3)** 16 h   **(4)** 19 h
   (5) none of these

3. 1 2 3 4 5

4. It will take 12 weeks and 6 days to do a job out in the desert. 3 work crews will work in the desert the same amount of time. How long will each crew spend out in the desert?
   (1) 4 weeks 1 day   **(2)** 5 weeks   **(3)** 4 weeks 2 days
   (4) 36 days   **(5)** none of these

4. 1 2 3 4 5

5. 3 people slept a total of 1 day and 3 hours. Each slept an equal amount of time. How long did each person sleep?
   (1) 8 h   **(2)** 10 h   **(3)** 7 h   **(4)** 9 h
   (5) none of these

5. 1 2 3 4 5

6. How many total seconds in 5 mn 32 seconds?
   (1) 332 s   **(2)** 300 s   **(3)** 302 s   **(4)** 320 s
   (5) none of these

6. 1 2 3 4 5

7. How many days in 144 hours?
   (1) 5   **(2)** 12   **(3)** 6   **(4)** 4
   (5) none of these

7. 1 2 3 4 5

8. Joe unloaded furniture for 2 hours 20 minutes at work. He worked a total of 8 hours at his job. How much time was spent doing other things?
   (1) 5 h   **(2)** 5 h 40 min   **(3)** 4 h 50 min   **(4)** 6 h
   (5) none of these

8. 1 2 3 4 5

Check your answers. If all your answers are right, start the evaluation. If one or two are wrong, make sure that you understand the right answers, and then start the evaluation. If you made more than two mistakes, you should read Chapter 26 again before starting the evaluation.

8. ANSWERS

2.1. **(a)** 3 min   **(b)** 780 s   **(c)** 5 h   **(d)** 5 days   **(e)** 7 weeks

3.1. **(a)** 104 min = 1 h 44 min   **(b)** 6 days 6 h   **(c)** 3 weeks 1 day   **(d)** 5 h

4.1. **(a)** 3 min 10 s   **(b)** 2 min 49 s   **(c)** 1 h 24 min 45 s   **(d)** 4 years 1 week

5.1. **(a)** 135 min = 2 h 15 min **(b)** 4 weeks 4 days

(c) 4 weeks 10 days = 5 weeks 3 days **(d)** 1 day 16 min 48 s

6.1. **(a)** 2 h 6 s **(b)** 4 h 2 min **(c)** 3 weeks 1 day **(d)** 1 week 6 h

7.1. **(a)** 1 h 8 min **(b)** 9 min **(c)** 6 days **(d)** 1 year 4 weeks

(e) 480 s **(f)** 20 weeks 4 days **(g)** 2 min 39 s

(h) 7 years 6 weeks **(i)** 2 h 5 min 6 s **(j)** 4 days 13 h

7.2. Word Problems

1. (2) 1 h 15 min
20 min
− 50 min
1 h 85 min = 2 h 25 min

2. (2) 40 min
− 25 min
15 min

3. (1) 1 week = 7 days
7 × 3 h = 21 h

4. (3) 4 weeks 2 days
3) 12 weeks 6 days

5. (4) 1 day 3 h = 24 + 3 = 27 h

$$\frac{9}{3) 27}$$

6. (1) 60 s = 1 min
60 × 5 = 300 s
300 + 32 = 332 s

7. (3) $\frac{6}{24) 144 \text{ h}}$

8. (2) 8 h   = 7 h 60 min
− 2 h 20 min = 2 h 20 min
5 h 40 min

Evaluation: Measurement of Length, Liquid, Time, Weight

Add, subtract, multiply, or divide. Reduce when necessary. Watch the signs.

(a)
```
    973 g
  - 251 g
```

(b)
```
  23 yd 11 in
 +  4 yd 10 in
```

(c)
```
  2 ft 7 in
×       6
```

(d)
```
   3 lb 4 oz
 + 9 lb 9 oz
```

(e) 3) 4 lb 2 oz

(f)
```
   2 gal 3 qt 1 c
 + 5 gal 2 qt 1 c
```

(g)
```
  22 h 10 min
×         6
```

(h)
```
  34 lb 10 oz
- 28 lb 13 oz
```

(i)
```
  2 qt 1 pt
×       7
```

(j)
```
  15 weeks 3 days
 - 9 weeks 5 days
```

(k) 8) 16 weeks

(l)
```
  3 qt
- 1 qt
```

(m)
```
  5 lb 11 oz
×        8
```

(n)
```
  12 h 45 mn
+ 15 h 28 mn
```

(o) 10) 10 yd 2 ft 6 in

(p)
```
  368 cm
- 209 cm
```

Change the units

(a) 12 in = _____ ft

(b) 3 ft = _____ yd

(c) 1 lb = _____ oz

(d) 1,000 g = _____ kg

(e) 1 qt = _____ pt

(f) 60 min = _____ h

(g) 60 s = _____ min

(h) 1 m = _____ cm

(i) 1,000 m = _____ km

(j) 8 liquid ounces = _____ c

(k) 1 gallon = _____ qt

(l) 2 c = _____ pt

(m) 24 h = _____ day

Word Problems

1. Candy weights are measured in:
   (1) minutes   (2) grams   (3) feet   (4) km
   (5) none of these

   1. 1   2   3   4   5

2. To say "Oh" it takes about 1:
   (1) yard   (2) gallon   (3) second   (4) day
   (5) none of these

   2. 1   2   3   4   5

289

3. An airplane lands on a runway. We measure the length of the runway in:
   (1) hours   (2) meters   (3) cups   (4) pounds
   (5) none of these

3. 1 2 3 4 5

4. We measure the length of a pencil in:
   (1) ounces   (2) liters   (3) days   (4) centimeters
   (5) none of these

4. 1 2 3 4 5

5. A football player threw a 44 yard pass. How many feet is that?
   (1) 144 ft   (2) 132 ft   (3) 135 ft   (4) 148 ft
   (5) none of these

5. 1 2 3 4 5

6. A board 3 yds 2 ft 3 in is cut into 3 equal pieces. How long is each piece?
   (1) 1 yd 1 ft 1 in   (2) 2 ½ yd   (3) 1 yd 6 in
   (4) 1 yd 9 in   (5) none of these

6. 1 2 3 4 5

7. There are 252 inches of ribbon on a roll. How many feet of ribbon is that?
   (1) 21 ft   (2) 25 ft   (3) 23 ft   (4) 24 ft
   (5) none of these

7. 1 2 3 4 5

8. Tony washed 12 lbs 8 oz of dark clothes and 6 lbs 10 ounces of light clothes. How many more pounds of dark clothes did Tony wash?
   (1) 6 lb 4 oz   (2) 5 lb 14 oz   (3) 5 lb 2 oz
   (4) 6 lb   (5) none of these

8. 1 2 3 4 5

9. How many cups of yogurt could you get from a pot that holds 2 quarts?
   (1) 8 cups   (2) 10 cups   (3) 6 cups   (4) 12 cups
   (5) none of these

9. 1 2 3 4 5

10. Jose had 3 pints of beer. Alfredo had 1 quart. How much more beer did Alfredo need to have the same amount as Jose?
    (1) 2 pt   (2) 1 pt   (3) 1 cup   (4) 3 pt
    (5) none of these

10. 1 2 3 4 5

Check your answers. If you made mistakes, make sure that you understand the right answers. Go back to any chapters you need to study again. Then go on to Chapter 21.

ANSWERS

Add, subtract, multiply, or divide.

(a) 722 g   (b) 27 yd 1 ft 9 in   (c) 5 yd 6 in   (d) 12 lb 13 oz

(e) 1 lb 6 oz   (f) 8 gal 1 qt 1 pt   (g) 5 days 13 h   (h) 5 lb 13 oz

(i) 4 gal 1 qt 1 pt   (j) 5 weeks 5 days   (k) 2 weeks   (l) 2 qt

(m) 45 lb 8 oz   (n) 1 day 4 h 13 min   (o) 1 yd 3 in   (p) 159 cm

Change the units

(a) 1 ft   (b) 1 yd   (c) 16 oz   (d) 1 kg   (e) 2 pt

(f) 1 h   (g) 1 min   (h) 100 cm   (i) 1 km   (j) 1 c

(k) 4 qt   (l) 1 pt   (m) 1 day

Word Problems

1. (2) grams   2. (3) seconds   3. (2) meters   4. (4) cm

5. (2)
$$\begin{array}{r} 44 \\ \times\ 3 \\ \hline 132\,\text{ft} \end{array}$$

6. (4)
$$\begin{array}{r} 1\,\text{yd}\qquad 9\,\text{in} \\ 3\,)\,\overline{3\,\text{yd}\ 2\,\text{ft}\ \ 3\,\text{in}} = \\ 3\,\text{yd}\qquad 27\,\text{in} \end{array}$$

7. (1)
$$\begin{array}{r} 21 \\ 12\,)\,\overline{252} \\ 24 \\ \hline 12 \\ 12 \\ \hline \end{array}$$

8. (2)
$$\begin{array}{r} 12\,\text{lb}\ \ 8\,\text{oz} = 11\,\text{lb}\ 24\,\text{oz} \\ -\ 6\,\text{lb}\ 10\,\text{oz} = \ 6\,\text{lb}\ 10\,\text{oz} \\ \hline 5\,\text{lb}\ 14\,\text{oz} \end{array}$$

9. (1) 2 c = 1 pt
2 pt = 1 qt
4 c = 1 qt
2 × 4 = 8 c

10. (2) 1 qt = 2 pt
3 pt − 2 pt = 1 pt

TWENTY–SEVEN

Signed Numbers

1. FAHRENHEIT TEMPERATURES

At 8 P.M. the thermometer reads 7° Fahrenheit. Overnight the temperature drops 12 degrees. What is the morning temperature?

Look at the thermometer.

Start at + 7 degrees.

Count down 12 degrees on the thermometer.

You'll have to go below zero.

The temperature in the morning is –5°F.

1.1. In the morning, the thermometer read –15°F. By late afternoon it read +8°F. How many degrees did the temperature rise?

Start at –15.

Count up the thermometer. Stop counting when you get to +8.

You counted 23 degrees up.

The temperature rose 23 degrees.

1.2. Test Yourself

In the evening the temperature was –8°F. Overnight the temperature fell 9 degrees. What did the thermometer say in the morning?

Check your answer. If it is right, go to section 1.3. If it is wrong, or you don't know what to do, go to section 1.2A.

1.2A. Start at –8. Count down 9 degrees.

What does the thermometer say?

(Don't forget the sign.) `

1.3. Practice

Use the Fahrenheit thermometer to find the temperatures.

temperature (°F)	change (°F)	new temperature (°F)
+10	drops 8	+2
+4	drops 9	
–6	rises 9	
–12	rises 10	
+7	drops 12	
–15	drops 3	
–4	rises 6	
+15	drops 5	

Check your answers.

2. CELSIUS TEMPERATURES

Temperature readings on the celsius thermometer are done the same way.

Water freezes at 0°C.

Water boils at 100°C.

The normal temperature of the human body is 37°C.

It is –12°C. Overnight the temperature goes down 4 degrees. What is the temperature in the morning?

Start at –12.

Count down 4 degrees.

Your answer: –16°C.

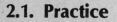

2.1. Practice

Use the Celsius thermometer to find the temperatures.

temperature (°C)	change (°C)	new temperature (°C)
+3	rises 8	+11
–5	rises 12	
–8	rises 15	
–14	drops 4	
+9	drops 12	
+5	drops 8	

3. HELPFUL HINTS FOR SIGNED NUMBERS

Signed numbers are like the degrees on a thermometer. Numbers above zero have + in front and are POSITIVE numbers. Numbers below zero have – in front and are NEGATIVE numbers. To work with signed numbers, draw a ladder and use it to count up or down.

4. ADDING A NUMBER TO A SIGNED NUMBER

To add, go up the ladder.
Example: +6 + 8
Start at +6.
Go up the ladder 8 steps.
Your answer: +14.

Example: –8 + 9
Start at –8.
Go up the ladder 9 steps.
Your answer: +1.

4.1. Test Yourself

How much is –9 + 3?
Check your answer. If it is right, go to section 4.2. If it is wrong or you don't know what to do, go to section 4.1A.
4.1A. To add, go up the ladder.
Start at –9.
Go up the ladder 3 steps.
What's the answer?

4.2. Practice

Use the ladder. Remember to put the sign in the answer.

    (a) +20 + 3 =   **(b)** –5 + 9 =   **(c)** –2 + 5 =   **(d)** +22 + 8 =

    (e) +7 + 6 =   **(f)** –7 + 12 =   **(g)** –10 + 10 =   **(h)** –9 + 15 =

Check your answers.

5. SUBTRACTING A NUMBER FROM A SIGNED NUMBER

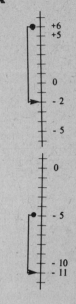

To subtract, go down the ladder.
Example: +6 – 8
Start at +6.
Go down the ladder 8 steps.
Your answer: –2.

Example: –5 – 6
Start at –5.
Go down the ladder 6 steps.
Your answer: –11.

5.1 Practice

(a) +26 – 13 =   **(b)** +3 – 8 =   **(c)** –2 – 7 =   **(d)** –6 – 8 =

(e) +4 – 8 =

Check your answers.

6. CANCELLING A SIGNED NUMBER

Look at the ladder.
Start at +3.
If you go down three steps, you arrive at zero.
So +3 – 3 = 0.
–3 cancels +3.

Start at –7.
To arrive at zero, you must go up 7 steps.
So –7 + 7 = 0.
+7 cancels –7.
To cancel a signed number, write the same digit with the other sign:

+2 cancels –2   because +2 – 2 = 0
–9 cancels +9   because –9 + 9 = 0

6.1. Practice

For each number, find the number that cancels it:

    (a) +3   **(b)** –5   **(c)** +25   **(d)** –32   **(e)** +7

Check your answers.

7. SIGNED NUMBERS—REMINDERS

Numbers above zero are positive. They have + in front.
Numbers below zero are negative. They have – in front.
Use a ladder to work with signed numbers.
To add, count up the ladder.
To subtract, count down.
Remember the + or the – in your answer.
To cancel a signed number, take the same digit with the other sign:

      +3 cancels –3   –7 cancels +7

7.1. More Practice

   (a) +16 + 8 = **(b)** +25 + 5 = **(c)** +4 – 14 = **(d)** +3 – 10 = **(e)** –19 + 7 =

7.2. Word Problems

1. The temperature rose from –7°F to +11°F. How many degrees did the temperature rise?
   (1) 16°F   **(2)** 18°F   **(3)** 4°F   **(4)** 0°F
   (5) none of these

1. 1 2 3 4 5

2. It was +13°C. The temperature dropped 15°C. What was the new temperature?
   (1) −28°C   **(2)** +2°C   **(3)** –2°C   **(4)** +28
   (5) none of these

2. 1 2 3 4 5

3. What do you add to –9 to get 0?
   (1) +9   **(2)** +18   **(3)** 0   **(4)** –9
   (5) none of these

3. 1 2 3 4 5

4. Death Valley, California, is 282 feet below sea level (–282). Los Angeles is at sea level (0). Los Angeles is how many feet above Death Valley?
(1) +282 ft   **(2)** –185 ft   **(3)** +504 ft   **(4)** 0 ft
(5) none of these

4. 1 2 3 4 5

5. You were in debt 88 dollars (–88). You paid all the money back. How much do you owe now?
(1) $88   **(2)** $0   **(3)** $178   **(4)** –88
(5) none of these

5. 1 2 3 4 5

Check your answers. If all your answers are right, start Chapter 28. If one or two are wrong, make sure that you understand the right answers and then start Chapter 28. If you made more than two mistakes, you should read Chapter 27 again before starting Chapter 28.

8. ANSWERS

1.2. –17°F

1.3. New Temperature (°F)
+2
–5
+3
–2
–5
–18
+2
+10

2.1. New Temperature (°C)
+11
+7
+7
–18
–3
–3

4.1. –6

4.2. (a) +23   **(b)** +4   **(c)** +3   **(d)** +30   **(e)** +13   **(f)** +5

(g) 0   **(h)** +6

5.1. **(a)** +13   **(b)** –5   **(c)** –9   **(d)** –14   **(e)** –4

6.1. **(a)** –3   **(b)** +5   **(c)** –25   **(d)** +32   **(e)** –7

7.1. **(a)** +24   **(b)** +30   **(c)** –10   **(d)** –7   **(e)** –12

7.2. Word Problems

  1. (2) Start at –7 and count up to +11.

  2. (3) +13 – 15 = –2

  3. (1) –9 + 9 = 0

  4. (1) –282 + 282 = 0

  5. (2) –88 + 88 = 0

TWENTY–EIGHT

Algebra

1. WHAT IS ALGEBRA?

Algebra is the part of mathematics that uses boxes or letters to stand for missing numbers. When you find the missing number you put it where the box or letter was.

Here is an algebra problem:

$$7 + 2 = \square$$

You say: 7 plus 2 equals what?

To find out, add 7 and 2.

Test the answer by putting it where the box was.

Does the answer in the box make the two sides equal?

You got it right. Your answer is 9.

In algebra, the amounts on both sides of the equal sign must be the same.

Here is another algebra problem:

$$9 - 6 = x$$

You say: 9 minus 6 equals what? The x is the missing number.

To find out, subtract 6 from 9.

Test the answer by putting it where the letter was.

Does the answer make the two sides equal?

You got it right. Your answer is 3.

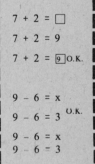

1.2. Test Yourself

$$15 - 7 = \square$$

Check your answer. If it is right, go to section 2. If it is wrong, or you don't know what to do, go to section 1.2A.

1.2A. You say: 15 minus 7 equals what?

To find out, subtract 7 from 15.

Then put your answer in the box to test it.

If it makes the two sides equal, you got it right.

2. HELPFUL HINTS FOR ALGEBRA

Each time you see a box or a letter in a problem, it is an algebra problem.

To find the solution, you must find what number should be in the place of the box or letter.

Test your answer by putting it in the box or in the place of the letter.

If you get the same amount on both sides, it's right.

You must get the same amount on both sides of the equal sign (=).

2.1. Practice

(a) $12 + 3 = \square$ (b) $50 + 30 = \square$ (c) $16 + 4 = P$

(d) $58 - 31 = \square$ (e) $47 - 23 = \square$

Check your answers.

3. ALGEBRA AND ADDITION

The box or letter can be in any part of the addition problem.

Example: $\square + 12 = 15$

You say: What plus 12 equals 15?

To find out, subtract 12 from both sides.

On the left side, +12 and –12 cancel each other.

The box stays by itself.

On the right side, you get 3.

Test the answer by putting it in the box.

$\square + 12 = 15$

$\square + 12 - 12 = 15 - 12$

$\square + 12 - 12 = 15 - 12$

     cancel   3

$\square = 3$

$\boxed{3} + 12 = 15$ O.K.

3.1. Test Yourself

$\square + 8 = 19$

Check your answer. If it is right, go to section 3.2. If it is wrong or you don't know what to do, go to section 3.1A.

3.1A. You say: What plus 8 equals 19?

To find out, you must cancel the 8. Subtract 8 from both sides.

This is what you get: $\square + 8 - 8 = 19 - 8$

Find the answer and test it by putting it in the box.

3.2. Practice

(a) $\square + 8 = 19$ **(b)** $14 + \square = 17$ **(c)** $25 + Z = 30$ **(d)** $50 + a = 58$

(e) $42 + x = 57$

Check your answers.

4. ALGEBRA AND SUBTRACTION

Example: $B - 7 = 2$
You say: What minus 7 is 2?
To find out, you must cancel the 7.
Add 7 to both sides.
On the left side, -7 and $+7$ cancel each other.
On the right side, you get 9.
Test the answer by putting it where the letter was.

$B - 7 = 2$
$B - 7 + 7 = 2 + 7$
$B - 7 + 7 = 2 + 7$
cancel   9
$B = 9$
$B - 7 = 2$
$9 - 7 = 2$ O.K.

4.1. Practice

(a) $\square - 14 = 4$ **(b)** $x - 5 = 20$ **(c)** $\square - 9 = 15$ **(d)** $a - 3 = 5$

(e) $\square - 17 = 3$

Check your answers.

4.2. If the box or letter is taken away, you must change the subtraction problem to an addition problem.

Example: $16 - Y = 12$

You say: 16 minus what equals 12?

The letter is subtracted, so you must change your problem to an addition problem.

To do that, add Y to both sides.
Cancel on left side.
Here is your addition problem:
Solve it by cancelling 12.
Your answer is 4.
Test it.

$$16 - Y = 12$$
$$16 - Y + Y = 12 + Y$$
$$16 - Y + Y = 12 + Y$$
$$\text{cancel}$$
$$16 = 12 + Y$$
$$16 - 12 = 12 + Y - 12$$
$$4 \qquad \text{cancel}$$
$$4 = Y$$
$$16 - Y = 12$$
$$16 - 4 = 12 \quad \text{O.K.}$$

4.3. Test Yourself

$6 - \square = 4$

Check your answer. If it is right, go to section 4.4. If it is wrong or you don't know what to do, go to section 4.3A.

4.3A. The box is subtracted, so you must change the subtraction problem to an addition problem.

To do that, add $\square$ to both sides. You get:

$$6 - \square + \square = 4 + \square$$
$$\text{cancel}$$

Your addition problem is:

$$6 = 4 + \square$$

You can solve it now (cancel the 4). Test your answer by putting it in the box.

4.4. Practice

(a) $9 - X = 5$ **(b)** $18 - V = 12$ **(c)** $28 - \square = 19$ **(d)** $50 - \square = 35$

(e) $34 - \square = 8$

Check your answers.

5. NUMBERS SENTENCES

In this chapter, we have been reading number sentences.

Examples:

$16 - Y = 12$ You say: 16 minus what equals 12?

$\square - 7 = 2$ You say: What minus 7 equals 2?

The box or the letter always stands for WHAT.

When you test your answer, you must get a true number sentence.

Examples:

$16 - 4 = 12$ You say: 16 minus 4 equals 12 (true)

$9 - 7 = 2$ You say: 9   minus 7 equals 2   (true)

5.1. Practice

Write the number sentences. WHAT must be replaced by a box or a letter.

(a) 4 plus 12 equals what?

(b) 9 minus what equals 2?

(c) 7 plus what equals 10?

(d) What minus 3 equals 15?

(e) 12 minus 6 equals what?

Check your answers.

6. CHANGING A WORD PROBLEM TO A NUMBER SENTENCE

Joe caught 6 fish in the morning. By late afternoon he had 15 fish in all. How many fish did he catch in the afternoon?

Here is what to do to make a number sentence with this problem.

15 is the total.

He caught 6 in the morning.

Then he caught more, so this will be an addition problem.

WHAT did he catch more? This is the missing number.

Put a box or a letter to show it.

Solve the problem and test the answer.

$$= 15$$
$$6 \quad = 15$$
$$6 + \quad = 15$$
$$6 + \square = 15$$
$$6 + \square - 6 = 15 - 6$$
$$= 9$$
$$6 + \boxed{9} = 15 \quad \text{O.K.}$$

6.1. Practice

Make a number sentence for each problem and find the answer.

(a) 48 people came to the meeting. 17 of them were male. How many were female?

(b) 19 fishing boats left the harbor before 4 A.M. Another 12 boats departed after 6 A.M. How many fishing boats went out in all?

(c) The radio announcer received 13 calls in 1 hour. 4 hours later a total of 48 people had called the station. How many people called the station after the first hour?

Check your answers.

7. ALGEBRA REMINDERS

Algebra is the part of mathematics that uses boxes or letters to show missing numbers.

$$5 + \square = 8 \qquad \text{This is a number sentence.}$$

When you solve the problem, you put the answer where the box or letter was. You must get a true number sentence.

$$5 + \boxed{3} = 8 \qquad \text{is true: } 3 \text{ is the right answer.}$$
$$5 + \boxed{2} = 8 \qquad \text{is not true: } 2 \text{ is a wrong answer.}$$

Read word problems carefully so you can change them to number sentences.

7.1. More Practice

(a) $24 - 8 = \square$ **(b)** $\square + 6 = 25$ **(c)** $\square - 12 = 68$ **(d)** $\square + 15 = 32$

(e) $42 + 12 = \square$ **(f)** $\square + 5 = 10$ **(g)** $14 - \square = 5$ **(h)** $19 - \square = 2$

Write the number sentence:

     (a) 8 minus 4 equals what?   **(d)** What minus 0 = 15?

     (b) What plus 16 equals 35?   **(e)** 16 plus what = 16?

     (c) 19 minus what equals 10?

Say if the number sentence is true or false

     (a) 43 + $\boxed{15}$ = 58   **(b)** 15 + $\boxed{7}$ = 8   **(c)** 9 − $\boxed{6}$ = 2   **(d)** 30 − 20 = $\boxed{10}$

     (e) $\boxed{5}$ − 8 = 23

Check your answers.

7.2. Word Problems

   1. Doris has some letters to mail. Clayton has 4 letters to mail. They have 9 letters to mail in all. How many letters does Doris have?
     (1) 13   **(2)** 5   **(3)** 9   **(4)** 4
     (5) none of these

     1. 1 2 3 4 5

   2. George brought home 18 mangoes. An hour later there were 14 mangoes left. How many mangoes did George eat?
     (1) 7   **(2)** 3   **(3)** 5   **(4)** 4
     (5) none of these

     2. 1 2 3 4 5

   3. 14 chairs were sold and there were 2 chairs left at the end of the sale. How many chairs were there at the beginning of the sale?
     (1) 16   **(2)** 12   **(3)** 15   **(4)** 18
     (5) none of these

     3. 1 2 3 4 5

   4. Ching is 4 years old. His age plus his mother's age equal 32. How old is Ching's mother?
     (1) 36   **(2)** 34   **(3)** 26   **(4)** 28
     (5) none of these

     4. 1 2 3 4 5

   5. What minus 12 equals 4?
     (1) 8   **(2)** 14   **(3)** 16   **(4)** 14
     (5) none of these

     5. 1 2 3 4 5

Check your answers. If all your answers are right, start Chapter 23. If one or two are wrong, make sure that you understand the right answers and then start Chapter 23. If you made more than two mistakes, you should read Chapter 22 again before starting Chapter 23.

8. ANSWERS

1.2. 8

2.1. (a) $\boxed{15}$ (b) $\boxed{80}$ (c) $P = \boxed{20}$ (d) $\boxed{27}$ (e) $\boxed{24}$

3.1. $\square = 11$

3.2. (a) $\boxed{11}$ (b) $\boxed{3}$ (c) $Z = 5$ (d) $a = 8$ (e) $x = 15$

4.1. (a) $\boxed{18}$ (b) $x = 25$ (c) $\boxed{24}$ (d) $a = 8$ (e) $\boxed{20}$

4.3. (a) $\boxed{2}$

4.4. (a) $x = 4$ (b) $V = 6$ (c) $\boxed{9}$ (d) $\boxed{15}$ (e) $\boxed{26}$

5.1. (a) $4 + 12 = \square$ (b) $9 - \square = 2$ (c) $7 + \square = 10$ (d) $\square - 3 = 5$

      (e) $12 - 6 = \square$

6.1. (a) $48 - 17 = \square$, $31 = \square$,   31 females

      (b) $19 + 2 = \square$, $31 = \square$,   31 fishing boats

      (c) $13 + \square = 48$, $13 + \square - 13 = 48 - 13$, $\square = 35$,   35 people called

7.1. (a) $\boxed{16}$ (b) $\boxed{19}$ (c) $\boxed{80}$ (d) $\boxed{17}$ (e) $\boxed{54}$ (f) $\boxed{5}$

      (g) $\boxed{9}$ (h) $\boxed{17}$

Write the number sentence:

      (a) $8 - 4 = \square$ (b) $\square + 16 = 35$ (c) $19 - \square = 10$ (d) $\square - 0 = 15$

      (e) $16 + \square = 16$

Say if the number sentence is true or false:

      (a) true   (b) false   (c) false   (d) true   (e) false

7.2. Word Problems

      1. (2) $4 + \square = 9$

            $4 + \square - 4 = 9 - 4$

                  $\square = 5$

            $4 + \boxed{5} = 9$ O.K.

      2. (4) $18 - \square = 14$

            $18 - \square + \square = 14 + \square$

                $18 = 14 + \square$

                  $4 = \square$

            $18 - \boxed{4} = 14$ O.K.

3. (1) $\boxed{} - 14 = 2$

$\boxed{} - 14 + 14 = 2 + 14$

$\boxed{} = 16$

$\boxed{16} = 2$ O.K.

4. (4) $4 + \boxed{} = 32$

$4 + \boxed{} - 4 = 32 - 4$

$\boxed{} = 28$

$4 + \boxed{28} = 32$ O.K.

5. (3) $\boxed{} - 12 = 4$

$\boxed{} - 12 + 12 = 4 + 12$

$\boxed{} = 16$

$\boxed{16} - 12 = 4$ O.K.

TWENTY–NINE

Graphs

1. WHAT IS A GRAPH?

A graph shows facts as a picture. If you learn how to read graphs, you'll be able to answer many questions about these facts.

2. PICTOGRAPHS

There are many types of graphs. Here is a pictograph:

Miles Driven By Salesperson

Each stands for 50 miles driven

In a pictograph, each picture represents a certain amount.

In this pictograph, each stands for 50 miles of driving.

So is 50 + 50, or 100 miles of driving.

2.1. How to Read a Pictograph

Read the title

Notice the label

Find the symbol for the information

Find what each symbol stands for.

You can find all the information you are looking for.

Which day had the most driving?

Find which day has the most ⊝

How many more miles were driven on Wednesday than on Friday?

> Title: Miles Driven by Salesperson
>
> Label: Days of the Week
>
> Symbol: ⊝
>
> Each ⊝ stands for 50 miles driven.
>
> Wednesday
>
> Wednesday: ⊝⊝⊝⊝ 200 miles
>
> Friday: ⊝⊝ 100 miles
>
> 100 more miles on Wednesday

2.2. Practice

Look at the pictograph to answer the questions.

(a) How many miles were driven all week?
(b) Which day had the least amount of driving?
(c) Which days did the salesperson drive 100 miles?

Check your answers.

3. BAR GRAPHS

Bar graphs show information using fat columns called bars. Bar graphs go horizontally (across) or vertically (up and down). Here is a horizontal bar graph:

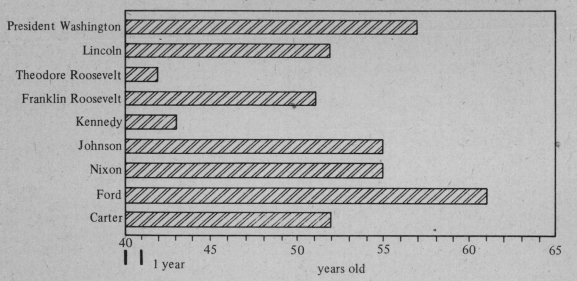

President's Ages at Inauguration

3.1. How to Read a Bar Graph

Read the title

Title: Presidents' Ages at Inauguration

Notice the labels

Label: years old, Presidents

Find the symbol for the information

Symbol: | |

Find what the symbol stands for

Each | | stands for 1 year

3.2. Practice

Look at the bar graph to answer the questions.

(a) How old was Kennedy at his inauguration?

(b) Who was the oldest President to be inaugurated?

(c) Which Presidents were 55 when they were inaugurated?

(d) How much older was Ford at his inauguration than Theodore Roosevelt was when he became President?

(e) How many Presidents were over 56 when they were inaugurated?

Check your answers.

3.3. Vertical Bar Graphs

Here is a vertical bar graph:

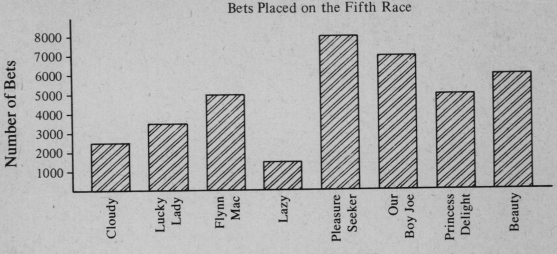

Bets Placed on the Fifth Race

Read the title	Bets Placed on the Fifth Race
Notice the labels	Names of Horses Number of bets
Find the symbol, scale, or key to information given	Bars of different lengths are shown
Determine the amount of the spaces	Begin at 1000, increase by 1000
Find the needed information. What horses have over 6000 bets? Find 6000 on the side. Go across and see which bars are above 6000.	Pleasure Seeker, Our Boy Joe
How many bets are on Lucky Lady? Find Lucky Lady on the bottom. The bar is between 3000 and 4000. That number is 3500.	3500

3.4. Practice

Use the graph to answer the questions.

(a) Which horse has the most bets (the favorite)?
(b) Which horse has the fewest bets?
(c) How many more bets are on Flynn Mac than on Cloudy?
(d) Lazy and Lucky Lady have a total of how many bets?
(e) Who is the third most popular horse?

Check your answers.

4. CIRCLE GRAPHS

Circle graphs, or pie graphs, show how something whole has been divided. The different parts into which something has been divided are shown like slices of a pie.

Circle graphs are excellent for showing where the money goes.

The Budget Dollar:
Where It Comes From

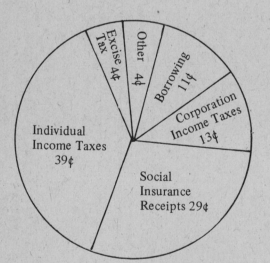

Read the title	The Budget Dollar: Where It Comes From
Notice the labels	Each slice tells where that part of the dollar goes
Find the symbol, scale, or key	The Budget Dollar—The pie should equal $1.

Find the information needed:

The greatest amount of money the federal government receives comes from _____.

Individual Income taxes—39¢

Look at the circle. Which fraction is the biggest?

8¢ of every dollar comes from _____. Look to see which slice or slices equals 8¢

4¢ excise taxes and 4¢ other

Individuals pay how much more per dollar than corporations in income taxes?

Look at the circle graph. Individuals pay 39¢

Corporations pay – 13

Individuals pay 26¢ more

$$\begin{array}{r} 39¢ \\ -\ 13 \\ \hline 26¢ \end{array}$$

4.1. A pie graph may use dots, stripes, or shaded areas to show divisions. Look for explanations of the sections

How People Get To Work In A Major U.S. City

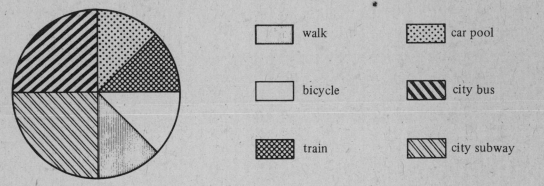

walk   car pool

bicycle   city bus

train   city subway

4.2. Practice

Use the graph to answer the questions.

(a) What two methods of transportation take up half of the graph?

(b) Bicycling and walking together take up what part of the graph?

(c) The number of people who take the bus is how much greater than the number of people who walk?

Check your answers.

5. LINE GRAPHS

A line graph shows changes in a quantity in relation to dates or amounts that don't change.

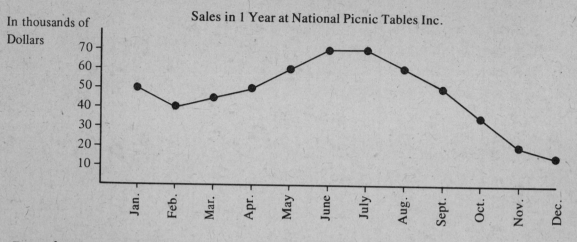

In thousands of Dollars

Sales in 1 Year at National Picnic Tables Inc.

5.1. Practice

(a) Which months had the same amount of sales?
(b) What was the amount of sales for May?
(c) What was the amount of sales for December?
(d) What was the total amount of sales for June and July?

Check your answers.

5.2. Two-line graphs show two changes. The lines may be different colors or one line may be solid and the other broken.

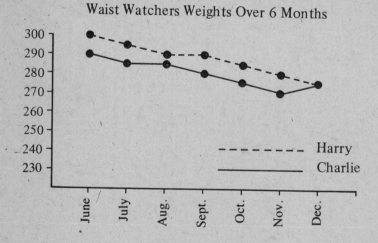

Waist Watchers Weights Over 6 Months

Before you answer the questions, look at Harry's entire record, then look at Charlie's

record. Make sure that you check [ ]
which line stands for Charlie.

5.3. Practice

Look at the graph to answer the following questions.
(a) How much did Charlie weigh at the beginning of the diet?
(b) How much did Charlie weigh after four months?
(c) Who lost more weight between July and August?
(d) Who lost the most weight in six months?
Check your answers.

6. GRAPHS REMINDERS

A graph is a way of giving facts about numbers or quantities by showing a picture.
When you read a graph, be sure to:
Read the title.
Notice the labels.
Find the symbol for the information.
Find what each symbol means.
Find the information you are looking for.

6.1. More Practice

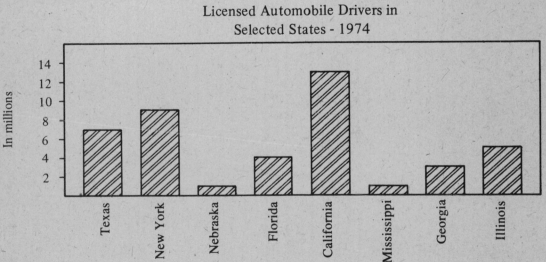

Licensed Automobile Drivers in
Selected States - 1974

(4) 0

2. C... ...ow many licensed drivers, in millions?
   (1) ...on   **(2)** 14 million   **(3)** 13 million
   (4) 12.5 million   **(5)** none of these

3. New York has how many more licensed drivers than Illinois?
   (1) 9 million   **(2)** 3 million   **(3)** 6 million
   (4) 4 million   **(5)** none of these

4. How many licensed drivers are there all together in Florida and Georgia?
   (1) 8 million   **(2)** 7 million   **(3)** 9 million   **(4)** 3 million
   (5) none of these

5. Which state has 7 million drivers?
   (1) Nebraska   **(2)** Florida   **(3)** Texas   **(4)** Georgia
   (5) none of these

A Farmer's Garden—1978

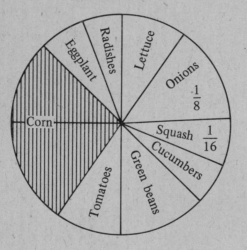

1. What part of the garden is for corn, onions and tomatoes?
   (1) ⅝   **(2)** ½   **(3)** ³⁄₁₆   **(4)** ¼
   (5) none of these

   1.   1   2   3   4   5

2. What part of the garden is for corn, squash, and lettuce?
   (1) ⅓   **(2)** ⁷⁄₁₆   **(3)** ¼   **(4)** ⅛
   (5) none of these

   2.   1   2   3   4   5

3. Eggplant and radishes take up the same amount of space as what other vegetable?
   (1) cucumbers (2) corn   (3) squash
   (4) corn and tomatoes   (5) none of these

   3. 1 2

4. The vegetable that takes the most part of the garden is
   (1) tomatoes (2) corn   (3) onions   (4) squash
   (5) none of these

   4. 1 2 3 4 5

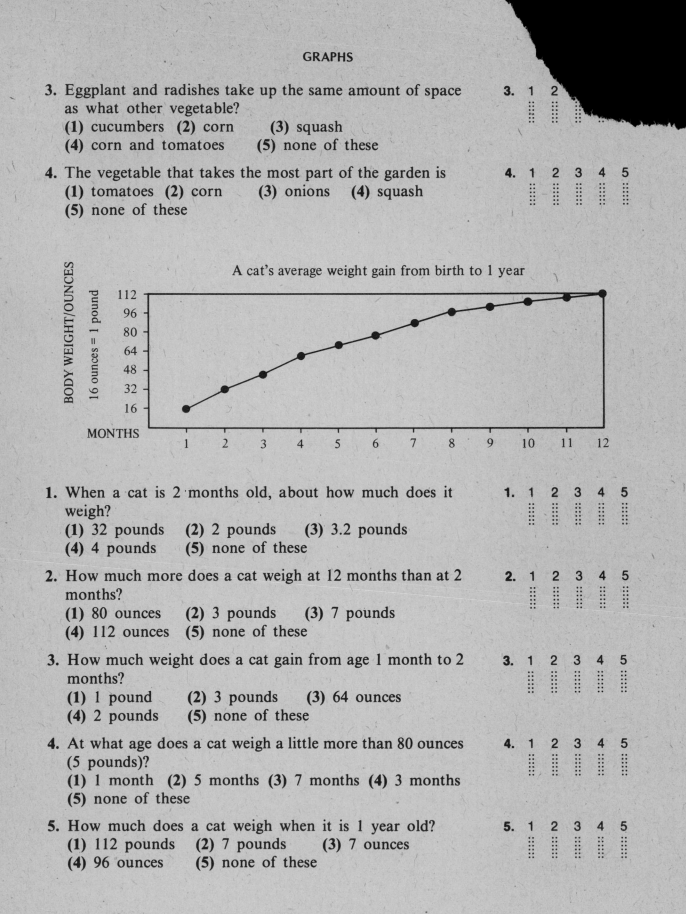

A cat's average weight gain from birth to 1 year

1. When a cat is 2 months old, about how much does it weigh?
   (1) 32 pounds   (2) 2 pounds   (3) 3.2 pounds
   (4) 4 pounds   (5) none of these

   1. 1 2 3 4 5

2. How much more does a cat weigh at 12 months than at 2 months?
   (1) 80 ounces   (2) 3 pounds   (3) 7 pounds
   (4) 112 ounces   (5) none of these

   2. 1 2 3 4 5

3. How much weight does a cat gain from age 1 month to 2 months?
   (1) 1 pound   (2) 3 pounds   (3) 64 ounces
   (4) 2 pounds   (5) none of these

   3. 1 2 3 4 5

4. At what age does a cat weigh a little more than 80 ounces (5 pounds)?
   (1) 1 month (2) 5 months (3) 7 months (4) 3 months
   (5) none of these

   4. 1 2 3 4 5

5. How much does a cat weigh when it is 1 year old?
   (1) 112 pounds   (2) 7 pounds   (3) 7 ounces
   (4) 96 ounces   (5) none of these

   5. 1 2 3 4 5

2.2 **(a)** 600 miles were driven all week

(b) Thursday   **(c)** Tuesday and Friday

3.2. **(a)** 43 years old   **(b)** Ford   **(c)** Johnson and Nixon

(d) 19 years older   **(e)** 2 (Ford and Washington)

3.4. **(a)** Pleasure Seeker   **(b)** Lazy   **(c)** 2,500   **(d)** 5,000 bets altogether

(e) Beauty

4.2. **(a)** city bus and city subway   **(b)** ¼   **(c)** ⅛ greater

5.1. **(a)** June and July; May and August; January, April, and September

(b) $60,000   **(c)** $15,000   **(d)** $140,000

5.3. **(a)** 290 pounds   **(b)** 275 pounds   **(c)** Harry   **(d)** Harry

6.1. BAR GRAPH

1. (2) 2 states

2. (3) 13 million

3. (4) 4 million (9 − 5 = 4)

4. (2) 7 million (4 + 3 = 7)

5. (3) Texas

CIRCLE GRAPH

1. (2) ½ (corn = ⅜, onions = ⅛, tomatoes = ⅛; ⁴⁄₈ = ½)

2. (2) ⁷⁄₁₆ (⅜ + ¹⁄₁₆ + ⅛ = ⁷⁄₁₆)

3. (5) none of these

4. (2) corn

LINE GRAPH

1. (2) 2 pounds

2. (1) 80 ounces

3. (1) 1 pound

4. (3) 7 months

5. (2) 7 pounds

s.

+23 – 23 =   **(d)** +16 – 19 =

(g) –76 + 76 =   **(h)** –14 + 22 =

(k) +52 – 43 =   **(l)** –8 – 9 =

(o) –9 – 10 =   **(p)** –20 + 21 =

(s) +5 – 19 =   **(t)** +6 – 6 =

or letter.

0   **(c)** 9 – □ = 0   **(d)** 24 – □ = 8

□ = 122   **(g)** 100 – 75 = p

6 = x **(j)** 49 + □ = 76

25 = 50   **(m)** □ + 15 = 90

□ = 7   **(p)** A + 37 = 100   **(q)** 58 + 42 = Q

+ 0 = 13   **(t)** W + 3 = 9

5 feet deep (–15). The boss said
ck into the hole (+10). Ground
t below ground is the hole?
  (3) –25   **(4)** –30

1. 1 2 3 4 5

rning. The temperature rose 12
ew temperature?
  (3) –5°C   **(4)** –19°C

2. 1 2 3 4 5

started out with 2 factories. Now
on. How many more factories has

  (3) 92   **(4)** 102

3. 1 2 3 4 5

line fo...
How muc...
(1) $40
(5) none of ...

5. Look at the grap...

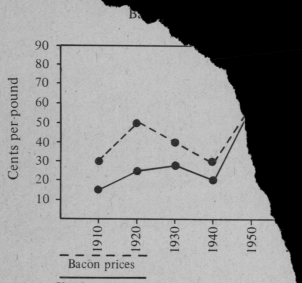

B...

Cents per-pound

90
80
70
60
50
40
30
20
10

1910 1920 1930 1940 1950

- - - Bacon prices

—— Chuck roast prices

What was the price of a pound of bacon in 1930?
(1) $.50   **(2)** $.30   **(3)** $.40   **(4)** $.35
(5) none of these

Chuck roast was how much less a pound than bacon
1920?
(1) $.25   **(2)** $.35   **(3)** $.50   **(4)** $.10
(5) none of these

How much did chuck roast jump in price between 194(
and 1950?
(1) $.30   **(2)** $.20   **(3)** $.60   **(4)** $.40
(5) none of these

In what year was chuck roast the same price as bacon?
(1) 1960   **(2)** 1950   **(3)** 1910   **(4)** 1930
(5) none of these

How much did bacon go down between 1920 and 1940?
(1) $.10   **(2)** $.15   **(3)** $.20   **(4)** $.05
(5) none of these

+20

4   **(f)** 8

2   **(l)** $x = 75$

$Q = 100$ **(r)** $w = 33$

These problems...
can. Check your ...
again.

1. $6 + 9 =$

5. $7 - 4 =$

6. $\begin{array}{r} 67 \\ - 21 \\ \hline \end{array}$

9. $6 \times 7 =$

10. $\begin{array}{r} 425 \\ \times\ \ 8 \\ \hline \end{array}$

13. $35 \div 5$

14. $7\ )\overline{916}$

17. $205\ )\overline{7380}$

18. $324\ )\overline{1,487}$

21. $\frac{5}{9} \div \frac{10}{18}$

22. $6\frac{2}{3} \div 5$

25. $\frac{13}{15} - \frac{3}{15}$

26. $6\frac{4}{5} - 2\frac{3}{5}$

29. $\begin{array}{r} 1\,\text{ft}\ \ 8\,\text{in} \\ + 2\,\text{ft}\ \ 9\,\text{in} \\ \hline \end{array}$

30. $\begin{array}{r} 5\,\text{yd}\ \ 1\,\text{ft}\ \ 4\,\text{in} \\ -\qquad\ \ 1\,\text{ft}\ \ 7\,\text{in} \\ \hline \end{array}$

31.

33. $\begin{array}{r} 1\,\text{gal}\ \ 2\,\text{qt} \\ +\qquad\ \ 3\,\text{qt} \\ \hline \end{array}$

34. $\begin{array}{r} 2\,\text{gal}\ \ 1\,\text{qt} \\ -\qquad\ \ 2\,\text{qt} \\ \hline \end{array}$

35.

37. $+3 - 5 =$

38. $-8 + 7 =$

39. 5

Read each problem carefully. Decide if you must ad...
sure that your answer fits the problem.

41. A couch is 438 dollars. A rug is 199 dollars, a roc...
is 79 dollars and 2 lamps cost 48 dollars. What i...
price of the living room furniture?
(1) $800   (2) $758   (3) $764   (4) $750
(5) none of these

42. 2,362 people paid grandstand admission to the stock car races. 6,896 people watched from the grass. How many more people were there on the grass than in the grandstand?
(1) 5,445   (2) 4,538   (3) 4,534   (4) 3,400
(5) none of these

43. 43 people bought color T.V.'s, 26 people had the sets delivered. How many people took the T.V. with them?
(1) 29   (2) 27   (3) 23   (4) 17
(5) none of these

42. 1 2 3 4 5
43. 1 2 3 4 5

44. The road maintenance crew paved 4 miles of highway every day for 15 days. There were how many miles of new paved road after 15 days?
(1) 80   (2) 60   (3) 50   (4) 59
(5) none of these

44. 1 2 3 4 5

45. There are 43 rows for parking at a shopping center. If 25 cars park in each row, how many cars can fit in the parking lot?
(1) 1,075   (2) 1,085   (3) 295   (4) 289
(5) none of these

45. 1 2 3 4 5

46. There are 64 chairs in a restaurant. How many tables are needed if there are to be 4 chairs at each table?
(1) 17   (2) 15   (3) 16   (4) 18
(5) none of these

46. 1 2 3 4 5

47. 12 people weigh a total of 2,220 pounds. Each person has the same weight. How much does each person weigh?
(1) 189   (2) 185   (3) 190   (4) 195
(5) none of these

47. 1 2 3 4 5

48. It takes $5\frac{1}{3}$ feet of lumber for a shelf. How many feet does it take for 3 shelves?
(1) $15\frac{2}{3}$ (2) 16   (3) 15   (4) $15\frac{1}{3}$
(5) none of these

48. 1 2 3 4 5

49. How many $\frac{1}{2}$'s are there in $6\frac{1}{2}$?
(1) 12   (2) 10   (3) 13   (4) 14
(5) none of these

49. 1 2 3 4 5

50. A large car has a $15\frac{1}{2}$ gallon gas tank. After driving 145 miles there are 6 gallons left. How many gallons did this gas guzzler use?
(1) 9   (2) $8\frac{1}{2}$ (3) $8\frac{3}{4}$ (4) $9\frac{1}{2}$
(5) none of these

50. 1 2 3 4 5

51. It takes 4⅓ cups of flour and 2½ cups of sugar for a cake. How many more cups of flour are there than sugar?
(1) 1⅚   (2) 2⅓   (3) 1½   (4) 1⅓
(5) none of these

51. 1 2 3 4 5

52. What is the total weight of 2 10-ounce cans of Cream of Chicken soup?
(1) 21 oz   (2) 1 lb 4 oz   (3) 1 lb 3 oz   (4) 1 lb 2 oz
(5) none of these

52. 1 2 3 4 5

53. It takes 25 minutes to sterilize a package of instruments in the operating room. How long would it take to serilize 9 packages?
(1) 3 hrs   (2) 4 hrs 10 mn   (3) 3 hrs 45 mn
(4) 3 hrs 10 mn   (5) none of these

53. 1 2 3 4 5

54. In the morning the thermometer said –7°F. By late afternoon it said +9°F. How many degrees did the temperature rise?
(1) 16°   (2) 0°   (3) 19°   (4) 2°
(5) none of these

54. 1 2 3 4 5

55.

WEATHER CHART

What is the difference between the temperature in May and the temperature in September?
(1) 50°   (2) 10°
(3) 60°   (4) 20°
(5) none of these

55. 1 2 3 4 5

ANSWERS

1. 15	**2.** 88	**3.** 1,177	**4.** 8,551
5. 3	**6.** 46	**7.** 384	**8.** 2,515
9. 42	**10.** 3,400	**11.** 2,451	**12.** 83,205
13. 7	**14.** 130r6	**15.** 20r5	**16.** 20r9
17. 36	**18.** 4r191	**19.** ⅔	**20.** 4

21. 1 **22.** $1\frac{1}{3}$ **23.** $\frac{4}{8} = \frac{1}{2}$ **24.** $7\frac{12}{9} = 8\frac{1}{3}$

25. $\frac{10}{15} = \frac{2}{3}$ **26.** $4\frac{1}{5}$ **27.** $13\frac{7}{9}$ **28.** $5\frac{5}{12}$

29. 3 ft 17 in = 1 yd 1 ft 5 in **30.** 4 yd 2 ft 9 in **31.** 10 lb 20 oz = 11 lb 4 oz

32. 1 lb 4 oz **33.** 2 gal 1 qt **34.** 1 gal 3 qt **35.** 1 week 1 day 20 h

36. 1 day 12 h **37.** −2 **38.** −1 **39.** $\boxed{14}$ **40.** $\boxed{5}$

41. (3)
$$\begin{array}{r} \$438 \\ 199 \\ 79 \\ +\ \ 48 \\ \hline \$764 \end{array}$$

42. (3)
$$\begin{array}{r} 6{,}896 \\ -\ 2{,}362 \\ \hline 4{,}534 \end{array}$$

43. (4)
$$\begin{array}{r} 43 \\ -\ 26 \\ \hline 17 \end{array}$$

44. (2)
$$\begin{array}{r} 15 \\ \times\ 4 \\ \hline 60 \end{array}$$

45. (1)
$$\begin{array}{r} 43 \\ \times\ 25 \\ \hline 215 \\ 860 \\ \hline 1{,}075 \end{array}$$

46. (3)
$$4\,\overline{)\,64}\quad = 16$$

47. (2)
$$\begin{array}{r} 185 \\ 12\,\overline{)\,2{,}220} \\ 1\,2 \\ \hline 1\,02 \\ 96 \\ \hline 60 \\ 60 \\ \hline \end{array}$$

48. (2) $5\frac{1}{3} \times 3 = \frac{16}{\cancel{3}_1} \times \frac{\cancel{3}^1}{1} = \frac{16}{1} = 16$

49. (3) $6\frac{1}{2} \div \frac{1}{2} =$

49. $\frac{13}{\cancel{2}_1} \times \frac{\cancel{2}^1}{1} = \frac{13}{1} = 13$

50. (4)　$15\frac{1}{2}$
　　　$-\ 6$
　　　——
　　　$9\frac{1}{2}$

51. (1)　$4\frac{1}{3} = 4\frac{2}{6} = 3\frac{8}{6}$
　　　$-\ 2\frac{1}{2} = 2\frac{3}{6} = 2\frac{3}{6}$
　　　————————
　　　　　　　　$1\frac{5}{6}$

52. (2)　10 oz
　　　$\times\ 2$
　　　———
　　　20 oz = 1 lb 4 oz

53. (3)　25 min　　　60 min = 1 h
　　　$\times\ 9$
　　　———　　　　　　　3 h 45 min
　　　225 min　　60) 225
　　　　　　　　　　　180
　　　　　　　　　　　———
　　　　　　　　　　　45

54. (1) –7 (count up to +9.) You count 16 steps
　　　16
　　　——

55. (2) 10°